MATH IN A CULTURAL CONTEXT©

Picking Berries

Connections Between Data Collection, Graphing, and Measuring

Part of the Series

Math in a Cultural Context:
Lessons Learned from Yup'ik Eskimo Elders

Grade 2
Also appropriate for Grade 3

Jerry Lipka

Janice Parmelee

Rebecca Adams

Developed at University of Alaska Fairbanks, Fairbanks, Alaska

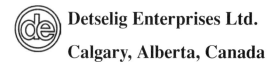 **Detselig Enterprises Ltd.**
Calgary, Alberta, Canada

Picking Berries: Connections Between Data Collection, Graphing, and Measuring

© 2004 University of Alaska Fairbanks

Library and Archives Canada Cataloguing in Publication

Lipka, Jerry

    Picking berries : connections between data collection, graphing and measuring / Jerry Lipka, Janice Parmelee, Rebecca Adams.

(Math in a cultural context: lessons learned from Yup'ik Eskimo elders)

Includes bibliographical references.

For use in grades 1–6.

ISBN 1-55059-282-3

    1. Mathematics—Study and teaching (Elementary) 2. Yupik Eskimos—Alaska. 3. Berries—Harvesting. I. Parmelee, Janice II. Adams, Rebecca III. Title. IV. Series.

QA135.6.L565 2004   372.7'044   C2004-906382-0

Math in a Cultural Context: Lessons Learned from Yup'ik Eskimo Elders© was developed at the University of Alaska Fairbanks. This material is based upon work supported by the National Science Foundation under grant #9618099, *Adapting Yup'ik Elders' Knowledge: Pre-K-to-6th Math and Instructional Materials Development,* and U.S. Department of Education grant #5356A030033, *Developing and Implementing Culturally Based Curriculum and Teacher Preparation.*

All rights reserved. No part of this book may be reproduced in any form or by any means without permission in writing from the publisher. The publisher grants limited reproduction permission for classroom teachers and aides for the purpose of making copies for overhead transparencies, student worksheets, handouts, student coloring book, and similarly related in-class student work.

 This project was sponsored, in part, by the National Science Foundation. Any opinions, findings, conclusions, or recommendations in this material are those of the author(s) and not necessarily those of the National Science Foundation (NSF) or the U.S. Department of Education.

 This project was also sponsored, in part, by the University of Alaska Fairbanks Alaska Schools Research Fund and the Bristol Bay Curriculum Project.

Detselig Enterprises Ltd. acknowledges the financial support of the Government of Canada through the Book Publishing Industry Development Program (BPIDP) for our publishing activities. We also acknowledge the support of the Alberta Foundation for the Arts for our publishing program.

Detselig Enterprises Ltd.
210-1220 Kensington Rd. N.W., Calgary, AB, T2N 3P5
Phone: (403) 283-0900/Fax: (403) 283-6947/E-mail: temeron@telusplanet.net
www.temerondetselig.com

ISBN: 1-55059-282-3

SAN: 113-0234

Printed in Canada.

MATH IN A CULTURAL CONTEXT©

Principal Investigator, Writer, and Series Editor:
Jerry Lipka

Project Mathematician:
Barbara Adams

Project Manager:
Flor Banks

Project Illustrator:
Putt (Elizabeth) Clark

Project Layout:
Sarah McGowan
Sue Mitchell

Curriculum Writers:
Claire Dominique Meierotto
Sandra Wildfeuer

Berry Picking Storybook Authors:
Stephen Walkie Charles
Agnes Green
Nastasia Wahlberg

Big John and Little Henry Storybook Author:
Seth Myers

Folklorist:
Ben Orr

Classroom Observers:
Randi Berlinger
Hannibal Grubis
Claire Dominique Meierotto
Sandi Pendergrast

Consultants:
Linda Brown
Richard Lehrer
Claire Dominique Meierotto
Clo Mingo
Ferdinand Sharp
Nastasia Wahlberg
Dan Watt
Peter Wiles
Evelyn Yanez

Teachers Piloting the Module:
Rebecca Adams
Barbara Arena
Darilynn Caston
Robert Childers
Nancy Douglas
Deborah Endicott
Frank Hendrickson
Sophie Kasayulie
Jannan Kaufman
Matthew Keener
Kelly Kelley
Seth Kelley
Elizabeth Lake
Sue Leatherbery
Don Long
Jackie Martin
Sheryl Martin
Paula McManus
Melissa Moede
John Nasset
Vanessa Nasset
T.J. O'Donnell
Sandi Pendergrast
Sassa Peterson
John Purcell
Nancy Sharp

Editing:
Iver Arnegard
Snježana Dananić
Katherine Mulcrone
Seth Myers
Kevin Peters
Nancy Slagle

Yup'ik Elders:
Mary Active
Henry Alakayak
Margaret Alakayak
Annie Blue
Lillie Gamechuk Pauk
Frederick George
Mary George
Samuel Ivan
George Moses
Anuska Nanalook
Anuska Petla
Wassily Simeon
Helen Toyukak
Margaret Wassillie
Evelyn Yanez

Yup'ik Translators:
Eliza Orr
Ferdinand Sharp
Nastasia Wahlberg
Evelyn Yanez

Table of Contents

Acknowledgements vii

Introduction: Math in a Cultural Context: Lessons Learned from Yup'ik Eskimo Elders ix

    Introduction to the Series. xi

    Introduction to the Picking Berries Module 1

Section 1: Getting Ready to Teach the Berries Module 21

Section 2: Organizing Discrete Data and Connecting it to Tables and Graphs

    Activity 1: Introduction to Berries 27

    Cultural Note: When Berry Picking Begins 30

    Activity 2: Describing Berries, Organizing Data, and Beginning Graphing Skills 36

        Part 1: Gathering Taste Data 36

        Part 2: Organizing Data to Create Graphs 37

Section 3: Collecting and Measuring Weather Data: Interpreting and Analyzing Data

    Activity 3: Building and Labeling a Graph 45

    Activity 4: Working with Graphs and Tables 51

    Activity 5: Building Graphs and Matching Tables 56

    Activity 6: Describing and Graphing Berry Data 60

    Cultural Note: Indicators of Seasonal Berry Supply 64

    Activity 7: Using Thermometers 66

        Part 1: Temperature and Thermometers 66

        Part 2: Temperature and Freezing Point 71

    Activity 8: Observing the Weather and Collecting Data 78

    Activity 9: Droopy Plants 82

    Activity 10: Where Do Berries Come From? 86

        Part 1: Map Studies 86

        Part 2: Using a Grid Map 88

        Part 3: Togiak Resource Mapping Game 92

Section 4: Refining Conceptual and Practical Understanding of Measuring and Manipulating Shadow Data

- Activity 11: Exploring Measurement 105
  - Part 1: Creating Units. 105
  - Part 2: Interpreting Height Data. 108
- Activity 12: Measuring Shadows 111
  - Part 1: Finding Common Units. 111
  - Part 2: Measuring Shadows and Answering Simple Conjectures 114
  - Part 3: Measuring Shadows with "Broken" Tape Measures 119
  - Part 4: Broken Rulers. 120
  - Part 5: Continuing to Answer Our Conjectures; Finding Patterns in Data 123
- Cultural Note: Uses of Berries 128
- Activity 13: Using the Berries and Preparing for the Berry Family Feast Day 131
  - Part 1: Feast Day Preparation 131
  - Part 2: Measuring and Cooking. 134
  - Part 3: Making *Akutaq* 136
- Activity 14: Reflections. 140

Acknowledgements

The supplemental math series *Math in a Cultural Context: Lessons Learned from Yup'ik Eskimo Elders,* is based on traditional and present-day wisdom and is dedicated to the late Mary George of Akiachak, Alaska. Mary contributed to every aspect of this long-term project.

This module, *Picking Berries: Connections Between Data Collection, Graphing, and Measuring,* is dedicated to the late Lillie Gamechuk Pauk of Manokotak, Alaska. Lillie was present at our very first meeting in Dillingham, Alaska, in the late 1980s when she shared with us how she measures. She opened our eyes to measuring without tools. She did not use a tape measure or any other instrumentation. Instead, she used her eyes to measure. This opened an important line of inquiry for us—non-Western ways of measuring accurately and efficiently. Also, Lillie taught us the importance of telling traditional Yup'ik stories. Even though she was ill upon arriving at one of our meetings, she chose to tell her story, and immediately after doing so she went into the hospital. Her actions and those of other elders underlined the importance of having a story to tell and their commitment to making these stories accessible to the next generation. We have come to include Yup'ik stories in most of these modules, in part because of the importance that elders like Lillie placed on them. Another outcome of this inquiry is this math module, which she inspired with her knowledge, her sense of humor, and her dedication. She made this module possible. Likewise, this work would not be possible without the assistance of many Yup'ik elders, community members, and teachers from Alaska.

For the past twenty-two years, I have had the pleasure to work with and learn from Evelyn Yanez of Togiak, Alaska; Nancy and Ferdinand Sharp and Anecia and Jonah Lomack of Manokotak, Alaska; Linda Brown of Ekwok and Fairbanks, Alaska; Sassa Peterson of Manokotak, Alaska; and Margie Hastings of New Stuyahok, Alaska. Their contributions are immeasurable, as is their friendship. My long-term relationship with elders who embraced this work wholeheartedly has made this difficult endeavor pleasurable as we learn from each other. In particular, I would like to acknowledge Henry Alakayak of Manokotak and Annie Blue of Togiak, whose dedication and commitment to cultural continuity has much inspired our project. They have set examples for our own perseverance creating culturally relevant school curricula. Equally unselfishly contributing were Mary Bavilla and Mary Active from Togiak and Anuska Nanalook from Manakotak—they came with stories that enriched us. Also, Sam Ivan of Akiak, Joshua Phillip and George Moses of Akiachak, and Anuska Nanalook and Anecia and Mike Toyukak of Manokotak provided knowledge about many aspects of traditional life, from Anecia's gifted storytelling and storyknifing to how kayaks were made and used and other traditional Yup'ik crafts to countless stories on how to survive.

I would like to acknowledge our dedicated staff, especially Flor Banks, with her highly refined organization skills, determination to get the job done, and motivation to move this project forward from reading and editing manuscripts to holding the various pieces of this project together; she has been an irreplaceable asset, and she has done it all with a smile. To Putt Clark, graphic artist extraordinaire, who kept up with every demand and produced more and better artwork than anyone could have hoped for and has worked with this project from its inception—thank you. To Barbara Adams for her clear-headed thinking and her mathematical insights that contributed so much depth to these modules and for her perseverance and dedication to completing a mathematically and culturally integrated math curriculum. To Dominique Claire Meierotto for her teacher insights and unselfish efforts creating, piloting, and helping with the writing and rewriting of this module. To Sandra Wildfeuer and Jan Parmelee, who wrote earlier drafts that laid the foundation for this module. A special thanks to Rebecca Adams for her energy and support during the summers of 2003 and 2004 that provided that extra push when we needed it. Thanks Becky! Also, Zana Dananic for her efforts during the summer of 2003 and through the 2003–2004 school year and summer of 2004 for editing, persisting, and keeping track of all those little details. I would like to thank Eliza and Ben Orr for for all their hard work and for producing the Yup'ik Glossary, an outstanding piece of work that continues to evolve and which accompanies this project. I wish to thank Walkie Charles, Nastasia Wahlberg, and Agnes Green who wrote the *Berry Picking* story.

I wish to acknowledge Kay Gilliland for her unflagging, hardworking dedication to this project. I wish to thank Richard Lehrer for his insights into constructivist approaches to measuring and for introducing us to Dan Watt. I would like to thank Dan for his critical review of different drafts of this module and for his knowledge, patience and attention to detail. A special thanks to Rebecca Adams, Sandi Pendergrast, Linda Brown, and Barbara Arena who dedicated the summer of 2003 in support of this work. Nancy Sharp stands out like no other. She has taught every module and her teaching has inspired many of us. I wish to thank John Purcell and Rob Childers, who went out of their way to teach the module, critique, and share their concerns and enthusiasm with us. Deborah Endicott, Vanessa Nasset, Frank Hendrickson, Nancy Sharp, Rebecca Adams, Matthew Keener, Kelly Kelley, Sue Leatherbery, Natalia Luehmann, Melissa Moede and Don Long allowed us into their classrooms and contributed the wonderful ideas to this project. Thanks to the Fairbanks North Star Borough School District, Lower Yukon School District, Yupiit School District, St. Mary's School District, Anchorage School District, Juneau School District, Yukon Flats School District, and Southwest Region Schools for their cooperation in piloting modules. And thanks to all the other math writers, project and pilot teachers, and elders who have assisted this project.

Thanks to Roger Norris-Tull, dean of the School of Education, University of Alaska Fairbanks, who supported this work generously. Sharon Nelson Barber, West Ed, has supported this work in spirit and in action for more than a decade. I wish to thank Sue Mitchell for her final editing and page layout work.

Last but not least, thank you to my loving wife, Janet Schichnes, who supported me in countless ways that allowed me to complete this work and to my loving children, Alan and Leah, who shared me with so many other people. The foundation of a loving and caring family provided the support and base to engage in this satisfying work.

Although this has been a long-term collaborative endeavor, I hope that we have taken a small step to meet the desires of the elders for the next generation so that they will be flexible thinkers able to effectively function in the Yup'ik and Western worlds.

Introduction

Math in a Cultural Context:

Lessons Learned from Yup'ik Eskimo Elders

Introduction to the Series

Math in a Cultural Context: Lessons Learned from Yup'ik Eskimo Elders is a supplemental math curriculum based on the traditional wisdom and practices of the Yup'ik Eskimo people of southwest Alaska. The kindergarten to sixth-grade math modules that you are about to teach are the result of more than a decade of collaboration between math educators, teachers, Yup'ik Eskimo elders, and educators to connect cultural knowledge to school mathematics. To understand the rich environment from which this curriculum came, imagine traveling on a snowmachine over the frozen tundra and finding your way based on the position of the stars in the night sky. Or, in summer paddling a sleek kayak across open waters shrouded in fog, yet knowing which way to travel toward land by the pattern of the waves. Or imagine building a kayak or making clothing and accurately sizing them by visualizing or using body measures. This is a small sample of the activities that modern Yup'ik people engage in. The mathematics embedded in these activities formed the basis for this series of supplemental math modules. Each module is independent and lasts from three to eight weeks.

From 2000 through spring 2003, with one exception, students who used these modules consistently outperformed at statistically significant levels over students who only used their regular math textbooks. This was true for urban as well as rural students, both Caucasian and Alaska Native. We believe that this supplemental curriculum will motivate your students and strengthen their mathematical understanding because of the engaging content, hands-on approach to problem solving, and the emphasis on mathematical communication. Further, these modules build on students' everyday experience and intuitive understandings, particularly in geometry, which is underrepresented in school.

The modules explore the everyday application of mathematical skills such as grouping, approximating, measuring, proportional thinking, informal geometry, and counting in base twenty and then present these in terms of formal mathematics. Students move from the concrete and applied to more formal and abstract math. The activities are designed to meet the following goals:

- Students learn to solve mathematical problems that support an in-depth understanding of mathematical concepts.
- Students derive mathematical formulas and rules from concrete and practical applications.
- Students become flexible thinkers because they learn that there is more than one method of solving a mathematical problem.
- Students learn to communicate and think mathematically while they demonstrate their understanding to peers.
- Students learn content across the curriculum, since the lessons comprise Yup'ik Eskimo culture, literacy, geography, and science.

Beyond meeting some of the content (mathematics) and process standards of the National Council of Teachers of Mathematics (2000), the curriculum design and its activities respond to the needs of diverse learners. Many activities are designed for group work. One of the strategies for using group work is to provide leadership opportunities to students who may not typically be placed in that role. Also, the modules tap into a wide array of intellectual abilities—practical, creative, and analytic. We assessed modules that were tested in rural Alaska, urban Alaska, and suburban California and found that students who were only peripherally involved in math became more active participants.

Students learn to reason mathematically by constructing models and analyzing practical tasks for their embedded mathematics. This enables them to generate and discover mathematical rules and formulas. In this way, we offer students a variety of ways to engage the math material through practical activity, spatial/visual learning, analytic thinking, and creative thinking. They are constantly encouraged to communicate mathematically by presenting their understandings while other students are encouraged to provide alternate solutions, strategies, and counter arguments. This process also strengthens their deductive reasoning.

Pedagogical Approach Used in the Modules

The curriculum design includes strategies that engage students:

- cognitively, so that students use a variety of thinking strategies (analytic, creative, and practical);
- socially, so that students with different social, cognitive, and mathematical skills use those strengths to lead and help solve mathematical problems;
- pedagogically, so that students explore mathematical concepts and communicate and learn to reason mathematically by demonstrating their understanding of the concepts; and
- practically, as students apply or investigate mathematics to solve problems from their daily lives.

The organization of the modules follows five distinct approaches to teaching and learning that converge into one system.

Expert-Apprentice Modeling

The first approach, expert-apprentice modeling, comes from Yup'ik elders and teachers and is supported by research in anthropology and education. Many lessons begin with the teacher (the expert) demonstrating a concept to the students (the apprentices). Following the theoretical position of the Russian psychologist Vygotsky (cited in Moll, 1990) and expert Yup'ik teachers (Lipka and Yanez, 1998) and elders, students begin to appropriate the knowledge of the teacher (who functions in the role of expert), as the teacher and the more adept apprentices help other students learn. This establishes a collaborative classroom setting in which student-to-student and student-to-teacher dialogues are part of the classroom fabric.

Reform-oriented Approach

The second pedagogical approach emphasizes student collaboration in solving "deeper" problems (Ma, 1999). This approach is supported by research in math classrooms and particularly by recent international studies (Stevenson et al., 1990; Stigler and Hiebert, 1998) strongly suggesting that math problems should be more in-depth and challenging and that students should understand the underlying principles, not merely use procedures competently. The modules present complex problems (two-step open-ended problems) that require students to think more deeply about mathematics.

Multiple Intelligences

Further, the modules tap into students' multiple intelligences. While some students may learn best from hands-on, real-world related problems, others may learn best when abstracting and deducing. This module provides opportunities to guide both modalities. Robert Sternberg's work (1997, 1998) influenced the development of these modules. He has consistently found that students who are taught so that they use their analytic, creative, and practical intelligences will outperform students who are taught using one modality, most often analytic. Thus, we have shaped our activities to engage students in this manner.

Introduction to the Series

Mathematical Argumentation and Deriving Rules

The purpose of math communication, argumentation, and conceptual understanding is to foster students' natural ability. These modules support a math classroom environment in which students explore the underlying mathematical rules as they solve problems. Through structured classroom communication, students will learn to work collaboratively in a problem-solving environment in which they learn both to appreciate alternative solutions and strategies and to evaluate these strategies and solutions. They will present their mathematical solutions to their peers. Through discrepancies in strategies and solutions, students will communicate with and help each other to understand their reasoning and mathematical decisions. Mathematical discussions are encouraged to strengthen students' mathematical and logical thinking as they share their findings. This requires classroom norms that support student communication, learning from errors, and viewing errors as an opportunity to learn rather than to criticize. The materials in the modules (see Materials section) constrain the possibilities, guide students in a particular direction, and increase their chances of understanding mathematical concepts. Students are given the opportunity to support their conceptual understanding by practicing it in the context of a particular problem.

Familiar and Unfamiliar Contexts Challenge Students' Thinking

By working in unfamiliar settings and facing new and challenging problems, students learn to think creatively. They gain confidence in their ability to solve both everyday problems and abstract mathematical questions, and their entire realm of knowledge and experience expands. Further, by making the familiar unfamiliar and by working on novel problems, students are encouraged to connect what they learn from one setting (everyday problems) with mathematics in another setting. For example, most sixth-grade students know about rectangles and how to calculate the area of a rectangle, but if you ask students to go outside and find the four corners of an eight-foot-by twelve-foot-rectangle without using rulers or similar instruments, they are faced with a challenging problem. As they work through this everyday application (which is needed to build any rectangular structure) and as they "prove" to their classmates that they do, in fact, have a rectangular base, they expand their knowledge of rectangles. In effect they must shift their thinking from considering rectangles as physical entities or as prototypical examples to understanding the salient properties of a rectangle. Similarly, everyday language, conceptions, and intuition may, in fact, be in the way of mathematical understanding and the precise meaning of mathematical terms. By treating familiar knowledge in unfamiliar ways, students explore and confront their own mathematical understandings and begin to understand the world of mathematics.

These major principles guide the overall pedagogical approach to the modules.

The Organization of the Modules

The curriculum comprises modules for kindergarten through sixth grade. Modules are divided into sections: activities, explorations, and exercises, with some variation between each module. Supplementary information is included in Cultural Notes, Teacher Notes, and Math Notes. Each module follows a particular cultural story line, and the mathematics connect directly to it. Some modules are designed around a children's story, and an illustrated text is included for the teacher to read to the class.

The module is a teacher's manual. It begins with a general overview of the activities ahead, an explanation of the math and pedagogy of the module, teaching suggestions, and a historical and cultural overview of the curriculum in general and of the specific module. Each activity includes a brief introductory statement, an estimated duration, goals, materials, any pre-class preparatory instructions for the teacher, and the procedures for the class to carry out the activity. Assessments are placed at various stages, both intermittently and at the end of activities.

Illustrations help to enliven the text. Yup'ik stories and games are interspersed and enrich the mathematics. Transparency masters, worksheet masters, assessments, and suggestions for additional materials are attached at the end of each activity. An overhead projector is necessary. Blackline masters that can be made into overhead transparencies are an important visual enhancement of the activities, stories, and games. Supplemental aids—colored posters, coloring books, and CD-ROMs—are attached separately or may be purchased elsewhere. Such visual aids also help to further classroom discussion and understanding.

Resources and Materials Required to Teach the Modules

Materials

The materials and tools limit the range of mathematical possibilities, guiding students' explorations so that they focus upon the intended purpose of the lesson. For example, in one module, latex sheets are used to explore concepts of topology. Students can manipulate the latex to the degree necessary to discover the mathematics of the various activities and apply the rules of topology.

For materials and learning tools that are more difficult to find or that are directly related to unique aspects of this curriculum, we provide detailed instructions for the teacher and students on how to make those tools. For example, in *Going to Egg Island: Adventures in Grouping and Place Values,* students use a base twenty abacus. Although the project has produced and makes available a few varieties of wooden abaci, detailed instructions are provided for the teacher and students on how to make a simple, inexpensive, and usable abacus with beads and pipe cleaners.

Each module and each activity lists all of the materials and learning tools necessary to carry it out. Some of the tools are expressly mathematical, such as interlocking centimeter cubes, abaci, and compasses. Others are particular to the given context of the problem, such as latex and black and white geometric pattern pieces. Many of the materials are items a teacher will probably have on hand, such as paper, markers, scissors, and rulers. Students learn to apply and manipulate the materials. The value of caring for the materials is underscored by the precepts of subsistence, which is based on processing raw materials and foods with maximum use and minimum waste. Periodically, we use food as part of an activity. In these instances, we encourage minimal waste.

Videos

To convey the knowledge of the elders underlying the entire curriculum more vividly, we have produced a few videos to accompany some of the modules. For example, the *Going to Egg Island: Adventures in Grouping and Place Values* module includes videos of Yup'ik elders demonstrating some traditional Yup'ik games. We also have footage and recordings of the ancient chants that accompanied these games. The videos are available on CD-ROM and are readily accessible for classroom use.

Yup'ik Language Glossary and Math Terms Glossary

To help teachers and students get a better feel for the Yup'ik language, its sounds, and the Yup'ik words used to describe mathematical concepts in this curriculum, we have developed a Yup'ik glossary on CD-ROM. Each word is recorded in digital form and can be played back in Yup'ik. The context of the word is provided, giving teachers and students a better sense of the Yup'ik concept, not just its Western "equivalent." Pictures and illustrations often accompany the word for additional clarification.

Values

There are many important Yup'ik values associated with each module. The elders counsel against waste. They value listening, learning, working hard, being cooperative, and passing knowledge on to others. These values are expressed in the contents of the Yup'ik stories that accompany the modules, in the cultural notes, and in various activities. Similarly, Yup'ik people as well as other traditional people continue to produce, build, and make crafts from raw materials. Students who engage in these modules also learn how to make simple mathematical tools fashioned around such themes as Yup'ik border patterns and building model kayaks, fish racks, and smokehouses. Students learn to appreciate and value other cultures.

Cultural Notes

Most of the mathematics used in the curriculum comes from our direct association and long-term collaboration with Yup'ik Eskimo elders and teachers. We have included many cultural notes to describe and explain more fully the purposes, origins, and variations associated with a particular traditional activity. Each module is based on a cultural activity and follows a Yup'ik cultural story line along which the activities and lessons unfold.

Math Notes

We want to ensure that teachers who may want to teach these modules but feel unsure of some of the mathematical concepts will feel supported by the Math Notes. These provide background material to help teachers better understand the mathematical concepts presented in the activities and exercises of each module. For example, in the *Perimeter and Area* module, the Math Notes give a detailed description of a rectangle and describe the geometric proofs one would apply to ascertain whether or not a shape is a rectangle. One module explores rectangular prisms and the geometry of three-dimensional objects; the Math Notes include information on the geometry of rectangular prisms, including proofs, to facilitate the instructional process. In every module, connections are made between the "formal math," its practical application, and the classroom strategies for teaching the math.

Teacher Notes

The main function of the Teacher Notes is to bring awareness to the key pedagogical aspects of the lesson. For example, they provide suggestions on how to facilitate students' mathematical understanding through classroom organization strategies, classroom communication, and ways of structuring lessons. Teacher Notes also make suggestions for ways of connecting out-of-school knowledge with schooling.

Assessment

Assessment and instruction are interrelated throughout the modules. Assessments are embedded within instructional activities, and teachers are encouraged to carefully observe, listen, and challenge their students' thinking. We call this active assessment, which allows teachers to assess how well students have learned to solve the mathematical and cultural problems introduced in a module.

Careful attention has been given to developing assessment techniques and tools that evaluate both the conceptual and procedural knowledge of students. We agree with Ma (1999) that having one type of knowledge without the other, or not understanding the link between the two, will produce only partial understanding. The goal here is to produce relational understanding in mathematics. Instruction and assessment have been developed and aligned to ensure that both types of knowledge are acquired; this has been accomplished using both traditional and alternative techniques.

The specific details and techniques for assessment (when applicable) are included within activities. The three main tools for collecting and using assessment data follow.

Journals

Each student can keep a journal for daily entries, consisting primarily of responses to specific activities. Student journals serve as a current record of their work and a long-term record of their increasing mathematical knowledge and ability to communicate this knowledge. Many of the modules and their activities require students to predict, sketch, define, explain, calculate, design, and solve problems. Often, students will be asked to revisit their responses after a series of activities so that they can appreciate and review what they have learned. Student journals also provide teachers with insight into students' thinking, making it an active tool in the assessment and instructional process.

Observation

Observing and listening to students lets teachers learn about the strategies that they use to analyze and solve various problems. Listening to informal conversations between students as they work cooperatively on problems provides further insight into their strategies. Through observation, teachers also learn about their students' attitudes toward mathematics and their skills in cooperating with others. Observation is an excellent way to link assessment with instruction.

Adaptive Instruction

The goal of the summary assessment in this curriculum is to adapt instruction to the skills and knowledge needed by a group of students. From reviewing journal notes to simply observing, teachers learn which mathematical processes their students are able to effectively use and which ones they need to practice more. Adaptive assessment and instruction complete the link between assessment and instruction.

An Introduction to the Land and Its People, Geography, and Climate

Flying over the largely uninhabited expanse of southwest Alaska on a dark winter morning, one looks down at a white landscape interspersed with trees, winding rivers, rolling hills, and mountains. One sees a handful of lights sprinkled here, a handful there. Half of Alaska's 600,000-plus population lives in Anchorage. The other half is dispersed among smaller cities such as Fairbanks and Juneau and among the over two hundred rural villages that are scattered across the state. Landing on the village airstrip, which is usually gravel and, in the winter, covered with smooth, hard-packed snow, one is taken to the village by either car or snowmachine. Hardly any villages or regional centers are connected to a road system. The major means of transportation between these communities is by small plane, boat, and snowmachine, depending on the season.

It is common for the school to be centrally located. Village roads are usually unpaved, and people drive cars, four-wheelers, and snowmachines. Houses are typically made from modern materials and have electricity and running water. Over the past twenty years, Alaska villages have undergone major changes, both technologically and culturally. Most now have television, a full phone system, modern water and sewage treatment facilities, an airport, and a small store. Some also have a restaurant, and a few even have a small hotel and taxicab service. Access to medical care and public safety are still sporadic, with the former usually provided by a local health care worker and a community health clinic, or by health care workers from larger cities or regional centers who visit on a regular basis. Serious medical emergencies require air evacuation to either Anchorage or Fairbanks.

The Schools

Years of work have gone into making education as accessible as possible to rural communities. Almost every village has an elementary school, and most have a high school. Some also have a higher education satellite facility, computer access to higher education courses, or options that enable students to earn college credits while in their respective home communities. Vocational education is taught in some of the high schools, and there are also special vocational education facilities in some villages. While English has become the dominant language throughout Alaska, many Yup'ik children in the villages of this region still learn Yup'ik at home.

Yup'ik Village Life Today

Most villagers continue to participate in the seasonal rounds of hunting, fishing, and gathering. Although many modern conveniences are located within the village, when one steps outside of its narrow bounds, one is immediately aware of one's vulnerability in this immense and unforgiving land, where one misstep can lead to disaster. Depending upon their location (coastal community, riverine, or interior), villagers hunt and gather the surrounding resources. These include sea mammals, fish, caribou, and many types of berries. The seasonal subsistence calendar illustrates which activities take place during the year (see Figure 1). Knowledgeable elders know how to cross rivers and find their way through ice fields, navigating the seemingly featureless tundra by using directional indicators such as frozen grass and the constellations in the night sky. All of this can mean the difference between life and death. In the summer, when this largely treeless, moss- and grass-covered plain thaws into a large swamp dotted with small lakes, the consequences of ignorance, carelessness, and inexperience can be just as devastating. Underwater hazards in the river, such as submerged logs, can capsize a boat, dumping the occupants into the cold, swift current. Overland travel is much more difficult during the warm months due to the marshy ground and many waterways, and one can easily become disoriented and get lost. The sea is also integral to life in this region and requires its own set of skills and specialized knowledge to be safely navigated.

The Importance of the Land: Hunting and Gathering

Basic subsistence skills include knowing how to read the sky to determine the weather and make appropriate travel plans, being able to read the land to find one's way, knowing how to build an emergency shelter and, in the greater scheme, how to hunt and gather food and properly process and store it. In addition, the byproducts of subsistence activities, such as carved walrus tusks, pelts, and skins are made into clothing or decorative items and a variety of other utilitarian arts and crafts products and provide an important source of cash for many rural residents.

Hunting and gathering are still of great importance in modern Yup'ik society. A young man's first seal hunt is celebrated; family members who normally live and work in one of the larger cities will often fly home to help when the salmon are running, and whole families still gather to go berry picking. The importance of hunting and gathering in daily life is further reflected in the legislative priorities expressed by rural residents in Alaska. These focus on such things as subsistence hunting regulations, fishing quotas, resource development, and environmental issues that affect the well-being of game animals and subsistence vegetation.

Conclusion

We developed this curriculum in a Yup'ik context. The traditional subsistence and other skills of the Yup'ik people incorporate spatial, geometrical, and proportional reasoning and other mathematical reasoning. We have attempted to offer you and your students a new way to approach and apply mathematics while also learning about Yup'ik culture. Our goal has been to present math as practical information that is inherent in everything we do. We hope your students will adopt and incorporate some of this knowledge and add it to the learning base.

We hope you and your students will benefit from the mathematics, culture, geography, and literature embedded in the *Math in a Cultural Context: Lessons Learned from Yup'ik Eskimo Elders* series. The elders who guided this work emphasized that the next generation of children should be flexible thinkers and leaders. In a small way, we hope that this curriculum guides you and your students along this path.

Tua-ii ingrutuq [This is not the end].

References

Lipka, Jerry, and Evelyn Yanez. (1998). "Identifying and Understanding Cultural Differences: Toward Culturally Based Pedagogy." In J. Lipka with G. Mohatt and the Ciulistet, *Transforming the Culture of Schools* (pp. 111–137). Mahwah, NJ: Lawrence Erlbaum.

Ma, L. (1999). *Knowing and Teaching Elementary Mathematics*. Mahwah, NJ: Lawrence Erlbaum.

Moll, L. (1990). *Vygotsky and Education: Instructional Implications and Applications of Sociohistorical Psychology*. Cambridge: Cambridge University Press.

National Council of Teachers of Mathematics. (2000). *Principles and Standards for School Mathematics*. Reston, VA: National Council of Teachers of Mathematics.

Sternberg, R. (1997). *Successful Intelligence*. New York: Plume.

Sternberg, R. (1998). "Principles of Teaching for Successful Intelligence." *Educational Psychologist* 33, 65–72.

Stevenson, H., M. Lummis, S.-Y. Lee, and J. Stigler. (1990). *Making the Grade in Mathematics*. Arlington, VA: National Council of Teachers of Mathematics.

Stigler, J., and J. Hiebert. (1998). "Teaching is a Cultural Activity." *American Educator* 22 (4), 4–11.

Introduction to the Series xix

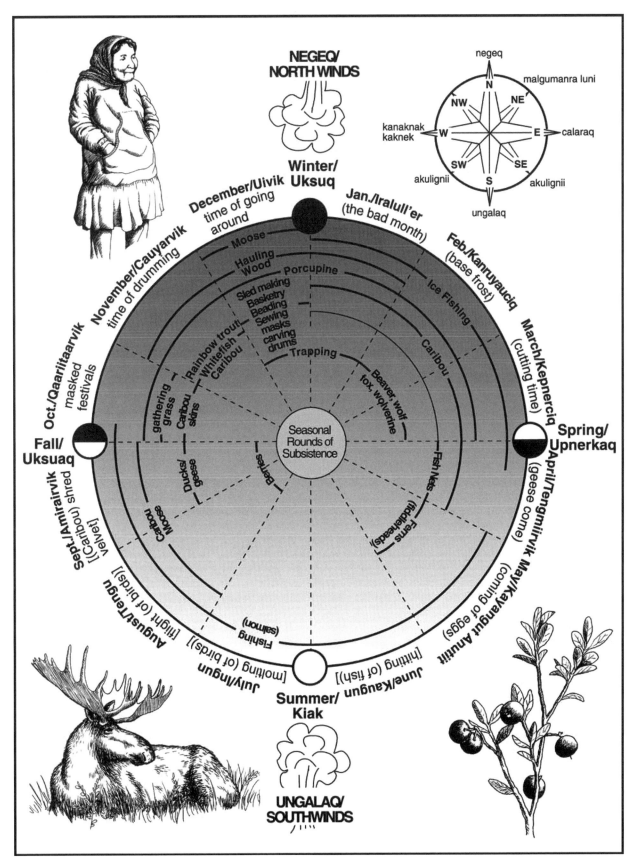

Fig. 1: Yearly Subsistence Calendar

Introduction to the Picking Berries Module

The series *Math in a Cultural Context* presents *Picking Berries: Connections Between Data Collection, Graphing, and Measuring*, a supplemental math module appropriate for second graders. This integrated curriculum connects aspects of Yup'ik culture to school mathematics and engages students in a series of hands-on activities that help them explore weather and seasonally related factors through math. This approach has been shown, statistically, to be effective in urban and rural Alaska and with varying ethnic groups such as Alaska Native, African American, and Caucasian.

An Overview of the Math

This module emphasizes both conceptual and applied understanding of measuring, data, and representation by engaging students in a series of activities that connect to everyday life. In this integrated approach to teaching mathematical concepts, the everyday phenomena of plant growth, in this case berries, connects student experiences to mathematical concepts. In this module, we connect the concept of a "unit" (an agreed-upon quantity or amount used for measuring) to measuring, data, and representation (graphing and tables). Students first generate data based on their descriptions of how a berry tastes. As the class works together to organize their unorganized data (the words used to describe how a berry tastes), they begin to understand how organized data can be interpreted and represented. The class eventually turns this data into bar graphs. As they make their graphs easier for others to read, they begin to understand the importance of labeling, using equal sized units (although not immediately required in graphing categorical data), and using values to represent the frequency of descriptive words. These and other concepts cycle throughout the module. Students represent data in both tabular and graphical forms, and they go back and forth from a table to create a graph and from a graph to create a table. A major milestone is accomplished once students recognize the relationship between tables and graphs. Students begin to see data as meaningful as they learn to ask and answer questions using their data.

In the next section of the module the students work on understanding how to read a thermometer, paying particular attention to increments. The mathematical milestone for students is to understand the concept of partitioning and partial units. Again, students work through conceptual understanding of partitioning and apply this to actual data. The students collect weather data integrating math and science and connect Yup'ik cultural knowledge and everyday experience to the math classroom.

Students measure their shadows as a way to connect the importance of sunlight to plant (berry) growth. In this section of the module, the students construct their own personal tape measure based on a nonstandard unit. As they develop their tape measure they face a host of problems that are directly related to understanding measuring. Specifically, they need to apply their unit consistently (iterating a unit), the unit needs to be equal sized (including units that do not overlap or leave gaps between), and how to partition (partial or fractional units) a unit.

Many students have difficulty knowing the difference between counting and measuring. The module draws connections between measuring and arithmetic by using number line strategies while measuring. Also, we have seen students not realize that measuring includes both the number and the unit. Students may say that Pete is taller than Mary because Pete is 10 units long while Mary is 5 units long. Students often do not realize the connection between the size of the unit and the number of times the unit is repeated. In this example Mary and Pete are actually the same size since the unit used to measure Mary is twice as long as the one used to measure Pete. Further, students may or may not realize the importance of starting and ending points while measuring.

We use the notion of a broken ruler and have students measure the length of their shadow or other objects with a broken ruler. Although students may show proficiency in using a standard ruler, when the ruler begins at 5 inches and ends at 10 inches, many students will report that a 5-inch-long object is 10 inches long because that is where the ruler ends.

Through these activities students work their way into understanding the importance of using common units or standard units and being able to communicate data and use this data in a meaningful way. To that end, students at the end of the module work with data, make conjectures, and collect data that responds to those conjectures. They learn how manipulate data and to interpret and analyze it. Again, the module is organized in a cyclical way so that by the end they can pose questions of data. Since they work with concrete data such as the length of a string that represents their shadow at a specific time, they can manipulate this data by answering questions such as who has the longest shadow in the class, or at what time of the day their shadow will be the shortest.

We developed this module around the subsistence activity of berry picking in rural southwest Alaska. The culture and math of the module connect through an engaging story of a family gathering berries during the end of summer and into early fall. This story motivates students to learn about math and the Yup'ik Eskimo culture and its values. Students also learn geography, practical problems, and factors that affect the berry supply as they listen to the story.

Berry picking is one of many traditional gathering activities described by Yup'ik Eskimo elders. Elders pay careful attention to many weather and seasonal variables that enable them to predict berry harvests—from the size of the harvest to the time when berries will be ready for harvest. Nastasia Wahlberg, who grew up in Bethel, the largest village in southwest Alaska, tells how people accomplish this important subsistence activity today and how it was done in the past in the Kuskokwim River area (see Figure 2). Even when Nastasia Wahlberg lived in Fairbanks, she still returned to Bethel every summer to pick berries.

Spatial abilities are used in this and other modules because they build on students' everyday experience. Further, spatial abilities and spatial tasks are designed to relate to geometrical relations and number sense. Thus, building on students' strength and background in spatial knowledge becomes a strategy for building and connecting to geometrical relations and number sense. For example, as students partition their own personal tape measure and create units, each equal unit divides space. As students represent this space numerically they are connecting spatial knowledge to numerical units. Thus, dividing space in equal increments or repeatedly adding equal units of space as a way to create a personal tape measure lays a foundation for multiplicative thinking. In the module, there are ample opportunities to build on these concepts through very practical activities such as adding and subtracting word problems that use a number line (personal tape measure).

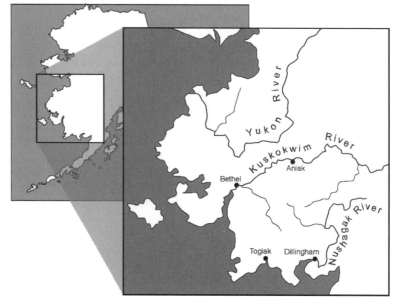

Fig. 2: The Kuskokwim Delta of Alaska

Why Use this Math Module?

Research Statistics Show Culturally Based Curriculum to be Effective

During the spring of 2003, we chose nine classrooms to pilot the *Picking Berries* module in rural and urban areas. We tested each of the classrooms before and after using the module. Groups using the culturally based math curriculum showed gains of 18.16% on average, while groups using standard math curricula gained an average of only 4.27% on the same exams. Over eighteen months, we tested this module in more than twenty classrooms and noticed a similar increase within the groups using the module.

Conceptual vs. Procedural Understanding of Measurement

Contemporary research on measuring by Stephan and Clements (2003), international assessments of students' math performance (National Center for Educational Statistics, 1996), and classroom research by Kamii and Clark (1997) all indicate that elementary students have difficulty mastering measurement concepts compared to other math topics. These researchers and others (Lehrer, 2003) indicate that there is as an overemphasis on procedural approaches to the teaching of measurement. In other words, too much time is spent on teaching students how to use rulers without developing a conceptual understanding of measurement.

Other researchers (Gravemeijer, 1998) suggest that elementary school curriculum underrepresents measurement and spatial reasoning. Yet, from their daily activities, children bring intuitive knowledge of spatial relations to school. As these researchers indicate, standard, procedurally oriented school math textbooks do not sufficiently build upon this intuitive understanding. Students should learn not only how to accomplish math-related tasks and procedures, but why they are done a certain way and why those ways work. A math curriculum needs to illustrate the reasoning behind the concepts.

One of the key strategies that this series of modules supports is working with students' and teachers' strengths as a way to build competence in other mathematical areas. Research and our own work in Alaska show that many rural community activities relate to spatial abilities, spatial manipulation, and spatial reasoning. For example, storyknifing, making patterns, traveling across the tundra or along rivers or open ocean, star navigating, and making clothing and other artifacts all rely on spatial abilities. Children in the community may or may not be directly engaged in such activities, but research shows that children can learn through this type of "intent participation" (Rogoff et al., 2003, pp. 175–176).

Students in general have not performed well on the measurement strands of international tests and, in particular, American Indian and Alaska Native students have lagged behind most other groups in general math performance. Thus, we have constructed a curricular approach to the teaching of this module that connects spatial sense to quantity. Our challenging, hands-on approach to the teaching of space and measurement appears to have succeeded in improving Alaska Native students' mathematical performance.

We use highly motivating activities and tasks that connect both to students' prior knowledge and to problems that are of immediate interest to them. The module challenges students to learn the

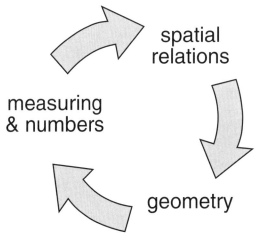

Fig. 3: Relation of mathematical concepts

underlying concepts of measuring: units, iteration, partitioning, and the meaning of markers and spaces. Our curriculum is novel, and thus inviting to study. The prospect of learning something new or something familiar presented in a new way draws students in.

Yup'ik Culture Informs Constructivist Learning Theory of Measurement and Data Collection

In general, traditional Yup'ik ways of teaching and learning support the practices emphasized and advocated by constructivist learning theory. The constructivist approach promotes active, hands-on learning in real-life contexts in which the teacher involves students in the process of learning and making meaningful connections between the curriculum and their life experiences. In the Yup'ik tradition, novices (children) often participate in tasks alongside experienced elders in order to learn the steps or processes necessary to arrive at the desired outcome, such as preparing a fish for drying.

Yup'ik people and others in a cross-cultural setting have shown us that there are various ways to perceive, conceive, and solve everyday math problems and to structure and measure space. Although the latest research (see *Learning and Teaching Measurement,* NCTM, Clements and Bright, editors, 2003) reports the need to structure space by iterating a unit, Yup'ik elders have their own method. Yup'ik adults are experts at measuring and solving a variety of problems associated with measuring (making their own clothes, traveling by star navigation, and building their own kayaks or other cultural artifacts). They rely on internal models developed from years of observations to develop their own units of measure, and may even mix units as they measure. People in traditional societies, carpenters, and crafts-oriented people often develop adaptive strategies for measuring. Measuring in such contexts is purposeful, since it accomplishes a particular task in a particular setting with available resources. Thus, they often view measuring relationally—what is the task, what are the resources, and who is the end user?

Another aspect of the elders' knowledge incorporated into this module is the stress on the importance of observing the weather and the environment, such as sunlight, precipitation, and temperature. Observing weather factors at a developmentally appropriate level is the central feature of this module. Students observe and record daily temperatures and plant growth and measure their shadows to observe changes in sunlight. These experiences teach students the procedures and basic meanings of measuring and data collection.

Math is Communication

Students as young as those in second grade may have already been socialized into expecting math to consist of "right" and "wrong" answers (Alrø and Skovsmose, 2002, p. 21). They may find it strange to "talk mathematically," and concepts such as proof may need to be slowly introduced. The students' justification of their answers is a step toward independent thinking.

Our constructive approach to mathematics requires the establishment of classroom social norms that promote a great deal of interaction and communication and encourage both students and teachers to view errors as opportunities to learn, not to criticize. These norms encourage students to use math vocabulary to express their reasoning and to gain mathematical knowledge. Through structured classroom communication and by encountering discrepancies in strategies and solutions, students learn to understand their mathematical reasoning and decisions.

Introduction to the Module

"Scaffolding" Supports Students as Independent Thinkers

One goal of this module, encouraged by elders such as Henry Alakayak, a long-time colleague and respected elder from Manokotak, is to guide students toward independent thinking: asking and answering their own questions by collecting and analyzing data. To do this, we have created an approach that "scaffolds" learning. The module begins by limiting the number of variables that the students have to deal with so they can explore mathematical concepts and conventions appropriate for second graders. After students have been supported with this "scaffolding," it is removed and they are challenged to proceed without support. Thus, we ease into the abstract concepts of starting points, units, unit iteration, and what the numbers on the ruler represent.

This module targets second grade students' accomplishment of the previously stated mathematical and pedagogical goals through an approach to problem solving in which the context of the lessons restrains and guides students' first steps toward independent thinking.

Integrating Literacy into the Math Classroom

The *Berry Picking* story is the primary literacy event that situates the math and personally connects students to western Alaska and the subsistence activities of the Yup'ik Eskimo people. This story exposes them to a different literary genre, since it has a different structure than students may be familiar with and teaches Yup'ik culture. Students also read *Big John and Little Henry,* a story that explores the development of standard units and the confusion that may occur from using nonstandard body measures.

The Math of the Module

Basic Organization

At the beginning of the module, students investigate discrete data and graphing. This investigation involves gathering data related to the berry harvest. Students organize this data into tables and graphs in order to understand the need for labels, values, and consistent units. Further, students learn various methods of representing the same data as tables and graphs. This occurs throughout the module.

As the module progresses, students explore the fundamental concepts of measurement. In order to understand these concepts and to illustrate the misunderstandings, students label and partition thermometers. Students also learn to make a line graph to represent data that changes over time.

Later in the module, students continue to work with measurement concepts by measuring the length (continuous data) of their shadows as a proxy for measuring sunlight. The activities related to this distance measurement focus on the basic concepts of units, the convenience of standard units, and the iteration of units that remain consistently sized and spaced.

The concepts involved in this module are interrelated and appear at various appropriate times throughout the activities. The mathematical concepts underlying a given activity are not independent, but constantly evolve and inform the other concepts in the module. For example, students collect weather data daily for use in multiple activities to illustrate multiple concepts.

Basic Concepts

Comparison

At a fundamental level, measurement is the comparison of two physical objects, numerically relating some measurement tool (a ruler, for example) with the object being measured. Basic comparison activities (see Figure 4) provide early success for students. Yet such activities can mask what students do not know about units and measurement.

As students develop their measurement sense, they use a third object, such as a ruler, as a proxy for an object that cannot be physically placed next to the object being compared. Students need a variety of experiences using standard and nonstandard measures to make such indirect comparisons.

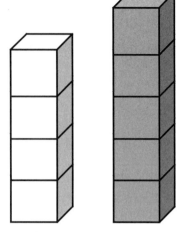

Fig. 4: How many more?

Data and Conjecture

The activities in this module begin the processes of understanding conjecture, explanation, and proof. We agree with the work of Curcio and Folkson (1996) asserting that "the success of child-generated tasks is based on two-criteria: (1) the information to be collected must be useful to the children, and (2) the information must be naturally interesting to them" (p. 385). We have tested our pedagogical methods for teaching these concepts and have found them effective in motivating students to generate conjectures.

Students who worked with their shadow lengths as a part of this module became excited and puzzled as they realized that their shadows move and change depending on the position of the sun. Students began to generate a series of conjectures and collected additional data to answer those conjectures. For example, "Will my shadow be longer than your shadow?" or "When will my shadow be the same length as my height?"

Fig. 5: Students' shadows

Graphing

Children in the early primary grades have had experiences representing data as pictographs and bar graphs, but they have less experience with line graphs and scatter-plot graphs. Different types of graphs can display the same data, but often one type of graph is better suited to certain data sets.

Figure 6 compares, on one grid, two different types of graphs for the same data set. Although each graph correctly represents the same data, the line graph is the most effective, since it can display a general trend of the changes over time.

Introduction to the Module

Categorical Graphs

Categorical graphs challenge students to define what is included and not included in a particular category. Categorical graphs can be created from discrete or continuous data. Students then classify data according their agreed-upon definitions. This provides a good opportunity for students to decide how to categorize data, and through this process students can build their vocabularies. Once the data is represented, comparisons can be made. Different types of data (discrete and continuous) necessitate different types of representations. Students' representations will show their current level of understanding. They may use both vertical and horizontal orientations for tables (and graphs), different-sized units, columns that drift across categories, and other "mistakes" that illustrate gaps in understanding. To remove the possibility of students' misunderstanding being masked, we have the students alter unit size and ask them to recreate categorical graphs in their journals.

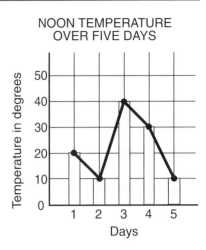

Fig. 6: Bar and line graphs for the same data set

After students accomplish the basic organization of a categorical graph, they label the axes, title the graph, and insert values. In choosing labels, titles, and scales, the students learn how to best represent data. Figure 7 illustrates the typical progress we have observed for second graders using this module.

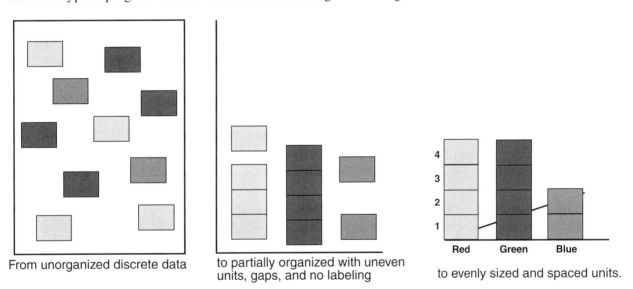

From unorganized discrete data | to partially organized with uneven units, gaps, and no labeling | to evenly sized and spaced units.

Fig. 7: Unorganized to organized

Bar Graphs

Bar graphs are often used to represent discrete and categorical data. They show distinct values and properties that a given category has, such as 'second-grade classroom in school X has 27 students,' or 'student A is five feet tall.' However, bar graphs can be also used to represent categorical data that is generally described as continuous. Figure 6 shows temperature values over five days that is a continuous type of data, yet the 'time of day' as a constant point allows for a bar graph representation.

Line Graphs

Continuous graphs display data that is uninterrupted in time or sequence. This type of information is usually portrayed as a line graph that may infer a trend or an average. However, a line does not always accurately portray the continuous data, since every pair of recorded values has other values between them. For example, Figure 6 shows only the noon temperatures during a period of five days, but not the continuous changes in daily temperature values based on each hour, each minute, or even each second. Another example of continuous data is seismographic data.

Scatter-plot Graphs

Data in a scatter-plot are represented as points that cannot necessarily be connected with a line. A scatter-plot displays points (x, y) on a grid that demonstrates the relationship of two variables. One may interpret the data in terms of the distribution of points, but the points do not bear a continuous relationship to each other (see Figure 8). Often in this case a linear regression line is calculated that summarizes the general trend in the graph. Calculating this regression line is beyond the scope of the module, but is an important idea for teachers to know.

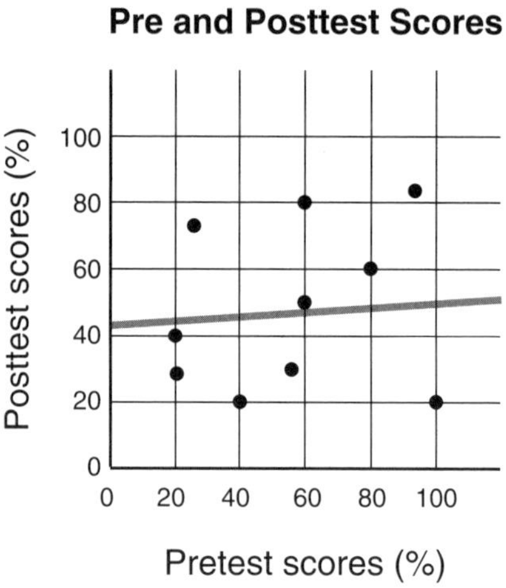

Fig. 8: Scatter-plot graph

Measuring and Counting

Piaget's research (Ginsburg and Opper, 1988, pp. 300–304) explains the connection between students' developmental level—specifically regarding conservation of length—and measuring. Piaget and his colleagues suggest, for example, that the concept of transitivity strongly relates to students' knowledge of measuring. More concretely, students need to be able to perform logical operations: if A = B and B > C, then A > C, for example. Students who do not understand this process may experience a variety of difficulties in measuring.

When students compare lengths measured with different units, the relative length of the object may be confused with the number of units counted. Unit size bears an inverse relationship to the value of the measurement: a smaller unit gives a larger count (value) while a larger unit gives a smaller count. In one teacher's class during the spring of 2003, students measured a certain length using small and large paperclips. Since the smaller paperclips resulted in a larger number of units than the larger paperclips did, these students reported that the distance measured with the smaller paperclip was longer (see Figure 9). This common error made by students provides great opportunity to discuss equivalence.

Fig. 9: Measuring using paperclips

Introduction to the Module

Standard and Nonstandard Measurement

This module includes the use of both standard and nonstandard measures and challenges students to understand some of the fundamental concepts associated with measuring. These concepts include the meaning of the marks and spaces on a measurement tool as well as partitioning, starting points, units, iterating units without gaps or overlaps, and partial units. At first, students use rulers and other standard measuring devices procedurally, that is, they read and record the values of lengths in inches or centimeters or in temperature scales such as Fahrenheit or Celsius. Students usually succeed in these tasks even though their successes may mask misunderstandings of the conceptual underpinnings of measurement. As the module progresses, the scaffolding of standard measurement is withdrawn, illustrating the limits of students' understanding. For example, a student may be able to say what the temperature is while misunderstanding what the hash marks on a thermometer represent.

Units and Partial Units

Students create personal tape measures incorporating chosen and iterated personal units in order to learn by comparison the efficiency of using standard systems of measurement. They learn the concepts of what a unit represents and measures, what the marks on a measurement tool (ruler/thermometer) mean, and how measurement is the accumulation of unit lengths. Understanding these concepts helps the students iterate a unit (repeat the unit without gaps or overlaps), partition a unit, and use partial units to begin to understand fractions.

However, these iteration tasks use only whole units; eventually students will have to measure an object whose length does not equal a whole unit. Students then solve the problem of partial units: how to represent them on their personal tape measures or thermometers and how to determine the value of the partial unit. During these activities, students learn that measurement is an approximation.

Assessment

We have given careful attention to developing assessment techniques and tools that evaluate both the procedural and conceptual knowledge of students. We agree with John A. Van de Walle (1998) that one type of knowledge without the other, or not understanding the link between the two, will produce only procedural understanding. Our goal is to produce conceptual understanding, allowing students to understand the how and why of mathematics and apply these concepts. Instruction and assessment, therefore, must be developed and aligned to ensure that students acquire both types of knowledge. We accomplish this by using both traditional and alternative assessment techniques.

Assessment activities are included in the lessons. These tasks also reinforce what students have learned. Assessment occurs as the students themselves explain and demonstrate what they have learned. This type of assessment directly relates to instruction and provides specific information for areas that require more practical applications. The main tools for collecting assessment data are listed below.

Assessment Tools

Daily Oral Language (Optional)

At the end of each day, students discuss what they have learned during their math lesson. Students agree on one or two of the main ideas and write sentences on the board describing those ideas for all to look over and agree on once again. After school, for the daily oral language lesson (DOL), insert errors into the sentences. Students will edit and rewrite while you check their work. When the sentences are correct, they will be the next entry in the students' math journals. Students may also write their sentences on their coloring book pages or write captions for their coloring book pages.

Journals

Journal writing is an effective way of teaching, reinforcing, and exploring math language and vocabulary, a central problem for math classes. The use of journals integrates features of the module to reinforce and connect the activities. In effect, journaling becomes a way to integrate different math concepts throughout the module. For example, as students reduce large data representations (such as the physical graphs and tables) to fit into their journals, they both interpret the data and scale it down.

Some teachers reported that journals were also an important way to inform students who were absent about what had occurred the day before. Journals are a way for students to review their activities and to improve their ability to retain information. The journals help many students understand the ongoing process of learning in math as well as spelling, grammar or usage, reading, and creative thinking.

Anecdotal Notes

You may choose to keep anecdotal notes on each student based on observations of student responses during group tasks or through individual monitoring. Specific criteria to help focus this note taking appear in each activity where anecdotal notes are indicated.

NCTM Standards and Math Goals of the Module: Principles and Standards for School Mathematics (2000)

Students in pre-kindergarten through grade 2 should be able to:

Number and Operations Standard
- Understand numbers, ways of representing numbers, relationships among numbers, and number systems:
    - Count with understanding and recognize "how many" in sets of objects
    - Understand and represent commonly used fractions, such as ¼, ½, and ¾
    - Connect number words and numerals to the quantities they represent, using various physical models and representations

Algebra Standard
- Analyze change in various contexts:
    - Describe qualitative changes, such as a student's growing taller
    - Describe quantitative change, such as a student's growing two inches taller
    - Sort, classify, and order objects by size, number, and other properties

Geometry Standard
- Specify locations and describe spatial relationships using coordinate geometry and other representational systems:
    - Describe, name, and interpret relative position in space and apply ideas about relative position
    - Describe, name, and interpret direction and distance in navigating space and apply ideas about direction and distance
    - Find and name locations with simple relationships such as "near to" and in coordinate systems such as maps

Measurement Standard
- Understand measurable attributes of objects and the units, systems, and process of measurement:
    - Recognize the attributes of length*
    - Compare and order objects according to these attributes
    - Understand how to measure using nonstandard and standard units
    - Select an appropriate unit and tool for the attribute being measured
- Apply appropriate techniques, tools, and formulas to determine measurements:
    - Measure with multiple copies of units of the same size, such as paper clips laid end to end
    - Use repetition of a single unit to measure something larger than the unit, for instance, measuring the length of a room with a single meterstick
    - Use tools to measure
    - Develop common referents for measures to make comparisons and estimates

Data Analysis and Probability
- Formulate questions that can be addressed with data and collect, organize, and display relevant data to answer them:
    - Pose questions and gather data about themselves and their surroundings
    - Sort and classify according to their attributes and organize data about the objects

- - Represent data using concrete objects, pictures, and graphs
- Select and use appropriate statistical methods to analyze data:
  - Describe parts of the data and the set of data as a whole to determine what the data show
- Develop and evaluate inferences and predictions that are based on data:
  - Discuss events related to students' experiences as likely or unlikely
- Understand and apply basic concepts of probability

Problem Solving Standard
- Build new mathematical knowledge through problem solving
- Solve problems that arise in mathematics
- Apply and adapt a variety of appropriate strategies to solve problems
- Monitor and reflect on the process of mathematical problem solving

Reasoning and Proof Standard
- Recognize reasoning and proof as fundamental aspects of mathematics
- Make and investigate mathematical conjectures
- Develop and evaluate mathematical arguments and proofs
- Select and use various types of reasoning and methods of proof

Communication Standard
- Organize and consolidate their mathematical thinking through communication, both verbal and nonverbal*
- Communicate their mathematical thinking coherently and clearly to peers, teachers, and others
- Analyze and evaluate the mathematical thinking and strategies of others
- Use the language of mathematics to express mathematical ideas precisely

Connections Standard
- Recognize and use connections among mathematical ideas
- Understand how mathematical ideas interconnect and build on one another to produce a coherent whole
- Recognize and apply mathematics in contexts outside of mathematics

Representation Standard
- Create and use representations to organize, record, and communicate mathematical ideas
- Select, apply, and translate among mathematical representations to solve problems
- Use representations to model and interpret physical, social, and mathematical phenomena

* modified from posted NCTM standards

> ## Alaska Standards for Culturally Responsive Schools (Alaska Native Knowledge Network, 1998)
>
> **A culturally responsive curriculum:**
> - reinforces the integrity of the cultural knowledge that students bring with them;
> - recognizes cultural knowledge as part of a living and constantly adapting system that is grounded in the past, but continues to grow through the present and into the future;
> - uses the local language and cultural knowledge as a foundation for the rest of the curriculum;
> - fosters a complementary relationship across knowledge derived from diverse knowledge systems; and
> - situates local knowledge and actions in a global context.

References

Alaska Native Knowledge Network. (1998). *Alaska Standards for Culturally Responsive Schools.* Fairbanks, AK: Alaska Native Knowledge Network.

Alrø, H., and Skovsmose, O. (2002). *Dialogue and Learning in Mathematics Education.* Boston, MA: Kluwer Academic Publishers.

Barett, J., and S. Dickson. (2003). "Understanding Children's Developing Strategies and Concepts of Length" in W. Bright and D. Clements (Eds.), *Learning and Teaching Measurement* (pp. 17–31). Reston, VA: National Council of Teachers of Mathematics.

Clement, J. (2001). Graphing. In R. Lehrer and L. Schauble (Eds.), *Investigating Real Data in the Classroom: Expanding Children's Understanding of Math and Science* (pp. 63–75). New York: Teachers College Press.

Clements, D., and G. Bright (Eds.). (2003). *Learning and Teaching Measurement.* Reston, VA: National Council of Teachers of Mathematics.

Clements, D., and M. Stephan. (2003). "Linear, Area, and Time Measurement in Prekindergarten to Grade 2" in G. Bright and D. Clements (Eds.), *Learning and Teaching Measurement* (pp. 3–17). Reston, VA: National Council of Teachers of Mathematics.

Curcio, F. R. (2001). *Developing Data-Graph Comprehension in Grades K–8.* Reston, VA: National Council of Teachers of Mathematics.

Curcio, F. R., and S. Folkson. (1996). Exploring Data: Kindergarten Children Do It Their Way. *Teaching Children Mathematics, 2,* 382–385.

Economopoulos, K., and T. Wright. (1999). *Investigations in Number, Data, and Space.* Menlo Park, CA: Dale Seymour.

Ginsburg, H. P., and S. Opper. (1988). *Piaget's Theory of Intellectual Development.* Englewood, NJ: Prentice Hall.

Gravemeijer, K. (1998). "From a Different Perspective: Building on Students' Informal Knowledge." In R. Lehrer and D. Chazan (Eds.), *Designing Learning Environments for Developing Understanding of Geometry and Space* (pp. 45–67). Mahwah, NJ: Lawrence Erlbaum.

Gravemeijer, K., J. Bowers, and M. Stephan (2003). A Hypothetical Learning Trajectory on Measurement and Flexible Arithmetic. *Journal for Research in Mathematics Education Monograph Number 12: Supporting Students' Development of Measuring Conceptions: Analyzing Students' Learning in Social Context.* (Ed. M. Stephan, J. Bowers, P. Cobb, and K. Gravemeijer).

Kamii, C., and F. B. Clark. (1997). "Measurement of Length: The Need for a Better Approach to Teaching." *School Science and Mathematics,* 97, 116–121.

Lehrer, R., and D. Chazan (Eds.). (1998). *Designing Learning Environments for Developing Understanding of Geometry and Space.* Mahwah, NJ: Lawrence Erlbaum.

Lehrer, R., et al. (2003). Developing an Understanding of Measurement in the Elementary Grades. In D. H. Clements and G. Bright (Eds.), *Learning and Teaching Measurement* (pp. 100–122). Reston, VA: National Council of Teachers of Mathematics.

Lehrer, R., and L. Schauble. (2003) Eds. *Investigating Real Data in the Classroom: Expanding Children's Understanding of Math and Science.* New York: Teachers College Press.

McDonald, M. (1997). *Tundra Mouse: A Storyknife Tale.* New York: Orchard Books.

National Center for Education Statistics. "Pursuing Excellence." Initial Findings from the Third International Mathematics and Science Study. NCES 97-198. Washington, D.C.: U.S. Government Printing Office, 1996. www.ed.gov/NCES/timss.

National Council of Teachers of Mathematics. (2002). *Navigating through Data Analysis and Probability in Prekindergarten–Grade 2.* Reston, VA: National Council of Teachers of Mathematics.

National Council of Teachers of Mathematics. (2000) *Principles and Standards for School Mathematics.* Reston, VA: National Council of Teachers of Mathematics.

Russell, S., R. Corwin, and K. Economopolous. (1997). *Does it Walk, Crawl, or Swim? Sorting and Classifying Data.* White Plains, NJ: Dale Seymour.

Rogoff, B., et al. (2003). Firsthand Learning through Intent Participation. *Annual Review of Psychology,* 54, 175–203.

Stephan, M., and D. H. Clements. (2003). Linear and Area Measurement in Prekindergarten to Grade 2. In Douglas H. Clements and George Bright (Eds.), *Learning and Teaching Measurement,* pp. 3–16. Reston, VA: National Council of Teachers of Mathematics.

Van de Walle, J. (1998). *Elementary and Middle School Mathematics: Teaching Developmentally.* Reading, MA: Addison-Wesley Longman.

Wainwright, S. (2001) Shadows. In R. Lehrer and L. Schauble (Eds.), *Investigating Real Data in the Classroom: Expanding Children's Understanding of Math and Science* (pp. 55–63). New York: Teachers College Press.

Resources

http://climate.gi.alaska.edu/weather/forecast/forecast_map.html
   Alaska Climate Research Center

http://illuminations.nctm.org
   An NCTM web site

http://nces.ed.gov/nceskids/graphing/index.asp
   National Center for Education Statistics

http://www.wrh.noaa.gov/wrhq/nwspage.html
   National Weather Service Home Pages

http://www.wunderground.com/US/
   To find the weather for any city, state, zip code or country

More Storybooks and Other Berry Related Books

Asch, Frank. (1985). *Bear Shadow.* Englewood Cliffs, NJ: Prentice-Hall.
    About a bear who tries to get rid of his shadow.
Beskow, E. (1998). *Peter in Blueberry Land.* Scranton, PA: Salem House.
    Peter meets the king of Blueberry Land as he looks for berries for his mother's birthday.
Bruchac, J. (1998). *The First Strawberries.* Madison: Turtleback Books.
    In this retelling of a Cherokee legend, a husband and wife quarrel until the Sun takes pity on the husband and sends berries to his wife, slowing her flight.
Burns, D. L. (2000). *Berries, Nuts, and Seeds.* Milwaukee: Gareth Stevens.
    This field guide contains information on berries, nuts, and seeds and how to identify the different species.
Degen, B. (1995). *Jamberry.* Madison: Turtleback Books.
Gardella, T. (2000). *Blackberry Booties.* New York: Orchard Books.
    Mikki Jo turns to her talent in picking blackberries after she is unsuccessful at making her new baby cousin a baby gift.
Goble, P. (1989). *Iktomi and the Berries: A Plains Indian Story.* Madison: Turtleback Books.
    Wonderful drawings illustrate this story of silly Iktomi trying to collect the reflection of buffalo berries in water.
Helman, A. (2003). *1 2 3 Moose: A Pacific Northwest Counting Book.* Seattle: Sasquatch Books.
    Color photographs and engaging text inform children about the ecology of the Pacific Northwest.
Hidaka, M. (1997). *Girl from the Snow Country.* La Jolla, CA: Kane/Miller Book.
    In northern Japan, a young girl uses red berries to decorate her snow bunnies.
Murphy, S. J. (1998). *A Fair Bear Share.* New York: HarperCollins.
    A mother bear will make her special blueberry pie if her cubs can gather enough seeds, nuts, and blueberries.

Master Materials List

Teacher Provides
Adding machine paper
Balls of string in two different colors
Bear Shadow by Frank Asch (children's storybook, optional)
Berries, fresh or frozen, as locally available
Butcher paper
Camera (optional)
Construction paper, 8½x11 and 11x17 inch
Containers for plants
Crayons, pencils, or markers
Cups, clear plastic
Glue sticks for each student
Graph paper
Ingredients for the *akutaq* recipe of your choice
Large bowl (optional)
Map, pull-down of the U.S. and Alaska (optional)
Maps, various geographical and political
Markers
Measuring cups or eye dropper
Napkins
Newspapers, scrap
Nonstandard measurement tools (paper clips, unifix cubes, string, books)
Paper, plain copy
Passin, or berry crusher (optional)
Pictures of Alaskan animals
Plants or plant seeds (lima beans, Wisconsin fast plants, marigolds, nasturtium, etc.)
Rain gauge (if available) or graduated cylinder
Resealable sandwich bags
Rulers
Safety scissors
Soil for plants
Stick-it note pads—½ (cut lengthwise) for each student
String, five-foot lengths
Student journals
Sugar, one bag
Tape, masking
Tape, transparent
Thermometers
    indoor/outdoor (one per class)
    small thermometers (one per pair or small group of students)
Utensils and bowls, enough for several "family groups"
Writing paper for student journals
Writing utensils

Introduction to the Module 17

Package Includes
Berries coloring book
Berry Picking storybook
Big John and Little Henry storybook
Yup'ik glossary CD [Teacher Reference]

Posters
Two included with the module:
   Berries
   Yup'ik Body Measures

Blackline Masters for Transparencies
Alaska
Berry Description Table (optional)
Crowberries
Blueberries
Kuskokwim River
Low Bush Cranberries
Map of the U.S. with Alaska Superimposed
North America
Salmonberries
Thermometer with Fahrenheit Increments of Ten
   Degrees
Thermometer with Fahrenheit Increments of Two
   Degrees
Togiak Region Map
Togiak Region Map with Grid
Togiak Resource Mapping Game
Togiak Resources Overlay

Blackline Masters for Handouts
Broken Rulers
Togiak Resource Mapping Game Cards

Blackline Masters for Worksheets
Alaska
Berry Description Table
Broken Ruler
Empty Thermometer
Horizontally and Vertically Oriented Temperature
   Tables
Making *Akutaq*
Map of the U.S. with Alaska Superimposed
Mary's Graph: Help Her Fix It
North America
Observing the Weather and Collecting Data
Pete's Graph: Help Him Fix It
Sample Recipes and T-tables
Thermometer with Fahrenheit Increments of Ten
   Degrees
Thermometer with Fahrenheit Increments of Two
   Degrees
Togiak Region Map
Togiak Region Map with Grid
Togiak Resource Mapping Game
Weather Table
Which Graph is Which?

Master Vocabulary List

Akutaq—known as "Eskimo Ice Cream," a traditional delight, or dessert, made of cloudberries, seal oil and sugar. Now it is often made with a variety of foods and usually begins with shortening.

Ascending Order—a way to organize data from the lowest to the highest number

Bar Graph—a graph that uses separate bars (rectangles) of different lengths to show and compare data

Base Line—the line on the graph where all data begins

Body Measures—a personalized or culturally prescribed way to measure distance using an individual's distinct body parts, such as the length of outstretched hands parallel to the ground

Cardinal Directions—north, south, east, west

Category—a way to organize and distinguish related objects

Categorical Graph—a graph representing discrete objects such as tastes or favorite colors

Columns—vertical data or lines

Common Unit of Measure—Any unit of measure, standard or nonstandard, chosen by a group as the agreed-upon unit of measure for a particular purpose or at a particular time

Compass Rose—a symbol on a map that shows direction

Conjecture—prediction based on incomplete evidence

Conservation of Length—length of an object remains the same even if the orientation of the object changes

Conservation of Area—area of an object remains the same if the orientation of the object has changed or even if it is decomposed into a set of pieces

Continuous Data—information that is provided by a phenomena that is uninterrupted in time or sequence, such as the change in temperature over time

Data—information

Data Interpretation—an explanation of a data set

Degree—a unit used when measuring temperature

Density—the amount of a substance contained within a specific length, area or volume

Descending Order—a way to organize data from the highest number to the lowest number

Discrete Data—data consisting of unconnected distinct values, such as the number of students in each classroom at a school

Estimating—the ability to determine any value, distance, size, or number

Freezing Point—the temperature at which water begins to freeze: 32 degrees Fahrenheit and 0 degrees Celsius.

Graph—a display of information through pictures or symbols on a coordinate system

Grid—a picture divided into equally spaced squares

Horizontal—something arranged in a side-to-side perspective

Iteration (unit iteration)—the action or process of repeating the same unit until a measure is achieved

Line Graph—a graph in which data points are connected by line segments

Introduction to the Module

Locating—finding the relative position of an object in space by using a coordinate system or relative position

Malruk naparneq—a Yup'ik length of measurement that is the same as two fists with thumbs extended and touching

Map Key—tells what the symbols on a map represent

Mercury—a silver-colored liquid that reacts quickly to heat and cold and is used in thermometers

Moisture—another word for water, and a term for water density

Natural Resource—plants, animals, or other materials that can be harvested for human use

Nonstandard Measure—a measure found by using a unit that differs from one person to another, such as using a body measure or paperclip instead of a ruler

Ordered Pair—a point on a graph or grid; the point represents both the value on the X- and Y-axes

Partial Unit—the amount of space left when measuring that is smaller than a whole unit

Partitioning—dividing space into equal-sized units

Passin—a traditional Native berry crusher

Proof—the validation of a proposition by application of specified rules; providing evidence to indicate that something is true, for example, $5 + 3 = 4 + 4$

Precipitation—any form of water or moisture that falls from the sky

Reflections—thoughts of personal experiences

Roots—the underground part of a plant that absorbs nutrients

Rows—horizontal data or lines

Scatter-plot—a graph made by plotting points on a coordinate plane to show the relationship between two variables in a data set. The points are not connected by line segments.

Shadow—the image cast by an object that is blocking a light source.

Soil—another word for dirt

Sorting—a process of organizing objects or data

Standard Measures—accepted ways of measuring using an established standard for example, inches or centimeters for length and pounds for weight

Starting Point—the point from which measuring begins; this is often referred to as the zero point on a ruler

Table—a display of data using rows and columns

Thermometer-- an instrument for measuring temperature, especially one having graduated measures along a glass tube filled with a liquid that expands as temperatures rise.

Transitivity—logical mathematical reasoning of the type: $A = B$, $B = C$, therefore $A = C$, or if $A > B$ and $B > C$ then $A > C$

Unit—An agreed-upon quantity or amount used for measuring, for example a foot as a unit of length.

Vertical—something arranged in an up-and-down perspective

Zero Point—the point where a measurement begins to count up units

Section 1:

Getting Ready to Teach the Berries Module

Getting Ready to Teach the Berries Module

Planning and preparation are important values to the Yup'ik people. Listed below are three exploration activities designed to prepare your students for future lessons within this module.

Knowing that supplies and classroom environments differ, as you read through these activities you will need to consider the best ways to adapt them to your teaching needs. You should also develop an appropriate timeline to initiate each of these explorations.

These explorations can best be accomplished by integrating them into your daily classroom schedule. You may also choose to expand and continue them beyond the scope of this module. For example, in the past, several teachers continued to gather weather and shadow data throughout the entire school year.

Exploration A: Droopy Plants

In this module we create a classroom proxy for how precipitation affects berry plant growth. We have students investigate how much water a plant needs. Ideally by having the students grow their plants from seeds they can also observe the plant cycle. If you will grow plants from seeds you should begin that process now. Other options include purchasing strawberry sprouts from a greenhouse, having students donate plants from home, or purchasing inexpensive plants. Under any option, be prepared for the classroom exploration with plants and water before Activity 9. For example, navy beans, marigolds, fast-growing seeds, and others all can be successfully grown in the classroom. For more detailed instructions refer to Activity 9, page 81.

Exploration B: Calendar Activities and Collecting Weather Data

Students can use outdoor thermometers or inexpensive plastic thermometers to record temperature at a set time each day. To include other weather data such as rainfall, students can use a graduated cylinder to provide them with an approximation of rainfall. For wind speed, develop a simple system of observing such as 0 to 10 miles an hour for when the leaves on trees move or when the smoke from a chimney goes

Fig. 10: Students record temperature

straight up. Work with the students to develop simple ways of gauging the data that they will collect. Collect between 5 and 12 days of data.

Students will collect weather data and analyze it. Wind, temperature, precipitation, and cloud coverage could be included in your daily calendar activities. Students may record these weather observations in their journals. You may have them glue an Observing the Weather and Collecting Data sheet into their journals to record their data on. We encourage you to have students collect this data over a long period of time so that they can compare weather information during different months and seasons. They will be able to observe trends and interpret data in a variety of ways. These daily activities will provide experience for data collection and interpretation included in Activity 8, page 78.

Section 2:

Organizing Discrete Data and Connecting it to Tables and Graphs

Activity 1:
Introduction to Berries

Your class will learn about a typical Yup'ik Eskimo food gathering activity—berry picking—and will read a storybook about this activity as well. This story introduces them to Yup'ik Eskimo culture and one important aspect of it—subsistence living—which in part relates to the gathering of food. As this cultural story develops, your students will learn some of the mathematics embedded in the cultural activity as well as classroom mathematics appropriate for the second grade. You can modify this module to make it appropriate for first and third graders.

In the next few lessons students will taste different berries, describe the taste, and sort descriptions into categories. They will collect discrete data (taste words), create a categorical graph without labels, and then label their graphs. Later in the module, students will describe berries using multiple attributes. They will also increase their graphing abilities as they understand that graphs need titles, labels, and values (numbers).

The first few lessons set the tone for the entire module. As students engage in the module's activities, try to promote a supportive and cooperative learning environment by encouraging guided student inquiry and student-to-student communication.

Students will begin by creating their math journals and math tool kits which they will use in their daily activities.

Goals
- To introduce the characters in the *Berry Picking* storybook
- To create covers for student math journals
- To organize a Math Tool Kit

Materials
- Berry coloring book (optional)
- *Berry Picking* storybook
- Poster, Berries
- Transparencies, Berries:
  Blueberries
  Low Bush Cranberries
  Crowberries
  Salmonberries
- Butcher paper

- Large (11x17 inch), light-colored construction paper for journal covers, one per student
- Paper for Student Journals
- Math Tool Kits (one per student):
  Crayons or pencils
  Glue sticks
  Colored fine-point markers or pens
  Resealable sandwich bags
  Stick-it note pads—(cut lengthwise)

Duration

One class period

Preparation

Place posters of Alaskan Berries in the room. Because the students will need pens, markers, and stick-it notes throughout the module, we suggest that the students create Math Tool Kits as a way to simplify classroom organization. Put out small resealable sandwich bags, colored fine-point markers, a pile of small stick-it notes cut in half, and anything else you might want on hand, such as glue sticks. Students line up to create their Math Tool Kits by putting one of each item in a resealable sandwich bag. They will keep the Math Tool Kits in their desks, ready to use throughout the module.

If your school is located near enough to berry patches and your teaching of this module coincides with berry-picking season, a walking field trip to gather berries is an excellent way to introduce this module. If possible, invite an elder to accompany the class on the field trip. Elders can enrich the module in numerous ways from story telling to their vast experience in understanding the factors that affect berry growth.

Instructions

1. Introduce the module by reading pages 1–5 of the *Berry Picking* storybook. Discuss the story, characters, and illustrations. What did you learn about the people and where they live? What does the mother tell Aanaq about weather and berries?

2. Show the different berry transparencies.

3. Ask students if they have collected any of these berries. Allow them to share a few stories of family outings involving berry picking.

4. Let them know that during this math unit, they will learn about berries, particularly about how Yup'ik Eskimo people of Alaska collect berries.

Activity 1: Introduction to Berries

5. Refer to one of the berry transparencies and ask, "Where do berries come from?" Students may say that they come from a store or from nature. Ask, "Where in nature might we find fresh berries? What do berry plants need to grow?" Take a few responses and write these on butcher paper for display.

6. Show the different berry transparencies again. Discuss the different ways that people use berries.

7. Have students take out their Math Tool Kits (see Preparation). Hand out construction paper for the journal covers. You may choose to have a variety of light-colored sheets available for students to choose from. Tell the students that this will become the cover for their math journals as they learn their lessons about berries.

8. Refer to the alternative suggestions below for making student journals. Choose or adapt the method of creating your student journals. Have students draw and color a picture on their berry journal cover and complete it with a journal title and student name. Extend this activity as you choose.

Alternative Journal Suggestions

Suggestion #1: Becky Adams, a teacher from North Pole, Alaska, found the following to be an effective approach for her students. Students chose a large (11x17 inch), light-colored sheet of construction paper and used markers to draw a full picture of their family picking berries. Beforehand, her students were told the paper would be folded in half to make their math journal covers. Students wrote their own title ideas and names on the outside covers, and made the inside front cover into a dedication page. Later, Adams folded and stapled these pictures to create the student journals. She used lined notebook paper in the front for daily oral practice (see page 9), plain copy paper in the middle, and more lined notebook paper near the end to form a glossary for student vocabulary development.

Suggestion #2: Debbie Endicott, a teacher from St. Mary's, Alaska, had her students write their own stories using the coloring book illustrations as a prompt. The students wrote creatively or connected their writing to what they learned in class.

Suggestion #3: Some teachers used the berries coloring book as a journal. The students made their journal entries directly into the coloring book, connecting the cultural storyline of picking berries with the math of the module.

Teacher Note

Nancy Sharp, a second-grade teacher in Manokotak, surprised her students by introducing this berry picking thematic unit. Nancy noted: "As I introduced the berry math module, some students looked at me as if to say, 'What in the world is she talking about when the berries are not growing yet?' Then when I started to read them the berries story, they reflected back to the memories of the times when they, too, went berry picking."

Cultural Note: When Berry Picking Begins

Gathering berries in rural southwest Alaska is a tradition and continues to be a means of subsistence for many Native Alaskans in rural areas today. Most of the berry picking begins in late July or early August, depending on the climate of the village. Nastasia Wahlberg of Bethel, our associate and consultant, mentions that temperature and humidity influence the speed at which the berries ripen. Weather greatly influences the picking season.

Fig. 1.1: Low bush cranberries

However, in other villages, the gathering of berries begins even sooner. *Tumagliq*—'low-bush cranberries' (*Vaccinum vitis-idaea*)—left on the tundra from the previous summer are best in early spring (see Fig 1.1). Anuska Petla, an elder who grew up in Koliganek and now lives in Dillingham, said that sometimes low bush cranberries left from the previous summer were "richer and juicier" than those of the current year.

Types of Berries

There are many types of berries that grow in southwest Alaska, and all the edible ones are picked and used by the Yup'ik Eskimos. The most plentiful varieties are listed below:

Yup'ik Name	English Name	Latin Name
atsalugpiaq (also called naunraq or aqevyik or aqevsik)	Cloudberry (often mistaken for salmonberries)	Rubus charmaemorus
curaq or suraq	blueberry	Vaccinium uliginosum
tan'gerpak	crowberry or curlewberry	Empetrum nigrum
tumagliq	low bush cranberry	Vaccinum vitis-idaea
mercuullugpak	high bush cranberry	Viburnum edule
uingiaraq	bog cranberry	Oxycoccus microcarplus
kavlak or kavlagpak	bearberry	Arctostaphylos
puyuraq or puyuraaq or puyurnaq	nagoonberry or wineberry	Rubus arcticus

People make separate trips to gather specific berries. A berry's particular season depends on its location and climate. In most areas blueberries (see Fig 1.2) and cloudberries are ready first, with crowberries and cranberries ripening later. Nastasia Wahlberg remembers gathering blueberries and cloudberries together, because they grow next to each other

Activity 1: Introduction to Berries 31

and are ripe at the same time, while crowberries and cranberries are best picked in August and September. Helen Toyukak, an elder from Manokotak, Alaska, adds:

> *[Crowberries] are best when the frost is on them. If they are picked too soon, they turn brown. We pick crowberries until they are covered with snow and we can't see them any more.*

As the time for berry picking approaches, small groups set out on short excursions to see how the berries are ripening. The scouts report to folks in the community. When the word is out that the berries are ready, trip preparations begin. Berries must be fairly ripe in order to soften and ripen further after they are picked.

Fig. 1.2: Blueberries

Nastiasia Wahlberg remembers, "In late July or early August, to get blueberries and cloudberries, we planned for a week-long excursion." However, not all people in southwest Alaska need to go on extended berry picking trips. Some live close enough to a good area to make it a day trip from home.

Fig: 1.3: Salmonberries

Fig. 1.4: Crowberries

Blueberries

Low Bush Cranberries

Crowberries

Blackline Master

Salmonberries

Blackline Master

Activity 2: Describing Berries, Organizing Data, and Beginning Graphing Skills

Research on math communication in math classrooms (Alrø and Skovsmose, 2002) shows that even preschool students believe that math is absolute, that there is only one correct answer and one way to get to it. The richness of mathematics occurs when the "right" answer isn't the only answer, when math is open-ended with multiple solution strategies. As a teacher of this module, you should take time to establish classroom norms that support student thinking and allow students to explain their thought strategies. Mutual math communication encourages the consensual understanding of concepts. In other words, teacher and student share the responsibility of knowledge production.

Consider the timeline you developed to begin the three exploration activities from page 23, Getting Ready to Teach the Module. You will now want to begin Exploration Activity B, collecting the weather data on a calendar. Although not stressed at this time, the calendar sheet acts as a data organizer much like a table. Each day and week contains weather information. Thus, the calendar doubles as a collection of data.

Part 1: Gathering Taste Data

In Part 1, students generate descriptive words of how berries taste to them. The students write these adjectives on stick-it notes and attach them to butcher paper on display for the whole classroom. You shouldn't organize this data, but allow it to appear in a disorganized form; this is raw data. When students begin to ask questions of data, they realize that for data to become meaningful it needs to be categorized and organized in a systematic way. The students decide how to organize this data, for example, by choosing whether words similar in meaning should be grouped together. Students learn the necessity of labeling their categories for purposes of mutual understanding.

As students copy the categorized data presented on the butcher paper into their journals, they realize the data cannot stay as big as it is; they must size it down to fit on journal paper. As students perform this task of scaling, they will intuitively use proportioning and begin to understand coding as a way to organize their data further, for example, using a smiling face to represent sweet or a frowning face to represent sour. These processes

Activity 2: Describing Berries 37

challenge students to think abstractly about practical information related to berries. Also, you and the students work together to make the graphs more understandable to outside observers.

Goals
- To create meaningful data from personal experience (descriptive words for how berries taste)
- To organize data meaningfully
- To pose and answer questions based on student data

Materials
- *Berry Picking* storybook
- Butcher paper
- Locally available berries, fresh or frozen (one berry per student)
- Math Tool Kits
- Student Journals
- Outdoor thermometer (one per class) or individual thermometer (one per two students)
- Worksheet, Observing the Weather and Collecting Data, one per student

Duration
One class period

Vocabulary
Data—information

Preparation
Cover a section of the board with butcher paper. Students will need their Math Tool Kits, their journals, and a berry for tasting. They will also need a copy of the Observing Weather and Collecting Data sheet.

Instructions
1. Read and discuss the *Berry Picking* storybook pages 7–11 and discuss how Aanaq knows it is time to pick berries. For additional information regarding bear grass, please read the Cultural Note: Indicators of the Seasonal Berry Supply on page 64.

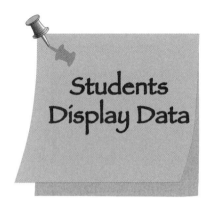

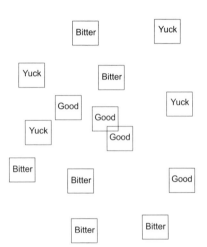

Fig. 2.1: Random display of berry taste words

2. You will begin collecting data based on Exploration B, Getting Ready to Teach the Module, found on page 23. Have students glue a copy of Observing the Weather and Collecting Data sheet into their journals. Students will transfer and record the pertinent daily information already begun on the classroom calendar to their worksheet. Students will add new information each day to this page. Students will need a new sheet at the end of the month in order to continue this data collection. See Exploration B on page 23 for details.

3. Give each student a berry (students will be working individually to begin this activity). Ask students to feel the weight of the berries, touch the outside, smell and look at them closely, allowing responses at each step. Now allow them to taste their berries.

4. Students will need markers and stick-it pads from their Math Tool Kits. Have each student write one word that best describes the taste of the berry and his or her name on a stick-it note.

5. Have each student come to the front of the room to stick his or her taste word onto the butcher paper (see Figure 2.1). Ask volunteers to share the word they chose and why they chose it.

6. Facilitate discussion as a large group. Ask, "Now that we have put our descriptive words on the board, what does this tell us?" At this point students may have a random collection of data that makes it very difficult to tell what the data means as a set. Ask, "What can we do to make the information more useful?" Discuss organizing the data.

7. Allow for student creativity and record their ideas at the side of the butcher paper.

Part 2: Organizing Data to Create Graphs

Part 2 involves the interaction between the three major math topics of this module: graphing, measuring, and data. The exercises also illustrate the underlying connection between measurement and numbers.

Students continue their work on categorical graphs and discrete data. They make the berry taste data easier to understand by reorganizing it into a graphic representation. They will most likely organize their data in columns or rows, horizontally or vertically. This is a step toward making a bar graph. Presenting data visually with a bar graph provides the opportunity to ask more specific questions of the information.

Activity 2: Describing Berries   39

Recording students' work and their dialogue provides an initial assessment for their progress in understanding these concepts. By returning to these classroom artifacts later in the module, you can gauge students' growth in understanding mathematical concepts relating to graphs and tables.

Goals
- To categorize information by similarities and differences
- To create a categorical graph (without numbers or labels)
- To generate questions based on their categorical data
- To realize that organized information improves communication and understanding of data
- To process information by organizing data

Materials
- Large construction paper, one per pair or small groups of students
- Descriptive word lists (teacher made from prior responses), one per pair or small group
- Math Tool Kits
- Student journals
- *Berry Picking* storybook

Duration
One class period

Vocabulary
Category—a way to organize and distinguish related objects
Horizontal—something arranged in a side-to-side perspective
Vertical—something organized in an up-and-down perspective

Preparation
Using yesterday's butcher paper display of descriptive words, write all the student word choices on one sheet of plain paper to be photocopied for each pair, or a small group of students. You or the students may cut the words apart so that the students can manipulate and organize their data.

Note: It is important for students to work through their organization of the data. Students vary in their abilities to organize the descriptive words. Some students may post these words randomly while others may display the words in an organized fashion—putting the berry tastes of one type together—so that they can see meaningful patterns. As students collectively solve problems like this, they will see the data set as meaningful.

Branching Out

This exercise is an opportunity for students to explore categories. Students create their own logically defined categories—what belongs and what does not. See Russell, S. et al. (1997) *Does It Walk, Crawl, or Swim: Sorting and Classifying Data,* White Plains, NY: Dale Seymour Publications. Also, visit NCTM's website for standard-based lessons on sorting and organizing data: http://illuminations.nctm.org.

Teacher Note

Figures 2.2 and 2.3 show two different data displays. Figure 2.2 shows a top-down orientation with columns clearly denoted. However, this student's use of space results in uneven-sized units. This may or may not be a problem, depending on the student's understanding of the display. This graph is without labels, titles, and values. Figure 2.3 uses circles to organize categories. The use of values indicates the number of items in each category. Challenge students to refine their displays further by imagining another person trying to understand what the data means.

Organized data allows students to read, interpret, and generalize. The data is then useful and meaningful.

Instructions

1. Read the *Berry Picking* storybook pages 13–17 and discuss the story. Ask the students: What the sisters, Aanaq and Kang'aq, do today? How do they get along?

2. Continue to record daily calendar and weather data into the student journals using the Observing the Weather and Collecting Data sheets.

3. Hand out construction paper and the photocopies of descriptive words (see Preparation above) to each pair or small group of students. Tell the students that they will make a picture (a graph) of the data they collected.

4. Allow the pairs or small groups to reorganize the descriptive words on their desktops to make the data more meaningful and useful. Students may combine similar words into one category, or create many categories.

Fig. 2.2: Strawberry tastes graph *Fig. 2.3: Strawberry tastes graph, uneven*

5. Once students have reorganized the data on their desks, have them share their organizational method with the class. Have the class visit each group's data display. Encourage student-to-student communication. This may be a good time to introduce the words "vertical" and "horizontal" in respect to various graphing methods used by the students. After all groups have reported, allow time for each group to make changes to their data display.

6. Have the groups glue their organized data onto large construction paper. Post these near the unorganized data display from Part 1 of the activity.

Activity 2: Describing Berries   *41*

7. Have your students make the data meaningful. Model, for the whole class, questions one could ask of the data and have students answer that question. For example, ask, "How many different descriptive words did we use? How many people wrote sweet? How many more or fewer people thought the berry tasted …?" Have the students ask questions of the data. Ask if their questions can be answered by looking at the data displayed in their graphs. Have students ask their questions to another member in their group and have each student answer the question. Have students write their questions in their journals and answer them. Allow time for students to share both their question and their answer, if they have one, with a friend.

8. Have students draw their group's graph in their journals.

Assessing Students' Developing Understanding of Graphs

Figure 2.4 is a graph from second-grade students graphing descriptive words. They show their graph as a pyramid. While observing the class we did not understand the students' representation. However, it appears that they have two orientations to their graph. There is a horizontal orientation for sweet, tasty, delicious, good, and sour, and a vertical orientation for liked. Further, the relative size of each "bar" represents the relative number of students who voted for that category. There are no individual units, just relatively sized blocks. The labeling can be followed but is difficult to read.

Yet, this representation provides an insight into how students "see" this data. A few lessons later this group began to organize their bar graphs into columns with a top-down or vertical orientation, to label each individual unit, and to place values at the top of each column. Units were still uneven and columns drifted. However, their later work could be much more easily read and understood. Students' journals provide valuable information concerning their growth and what they still do not understand.

Assessment Suggestions

Collect the journals and note how the students organized the data and what questions were generated and answered. Keep the students' first graphs displayed so they have a record of their learning. Make a master graph incorporating ideas from the entire class to display as a record as well.

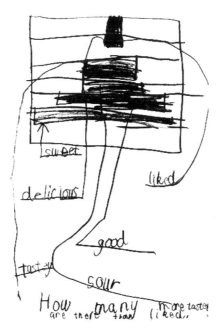

Fig. 2.4: Horizontal graph

Observing the Weather and Collecting Data

Sunday	Monday	Tuesday	Wednesday	Thursday	Friday	Saturday

Section 3:

Collecting and Measuring Weather Data: Interpreting and Analyzing Data

Activity 3:
Building and Labeling a Graph

The previous activity ended with students recreating the taste graph in their journals, a task that reveals students' intuitive understanding of discrete data. In this activity, students further develop the concepts of a categorical graph based on the same taste data.

In the previous activity, stick-it notes automatically determined unit size. In Activity 3, students learn that when applying units to a categorical graph, the units should be the same size and have no gaps between them. Also, the units should not overlap and they should start at a consistent starting point. Students learn these fundamental concepts of unit consistency, enabling them to measure by counting repeated units. This highlights the connection between measurement and number sense. In order to convey a measurement outside the context of its collection, units should be applied consistently.

In Activity 3, the class examines the graphs drawn in the students' journals, providing the opportunity for assessments of students' understanding. Some students will have drawn units of different sizes starting at different points. Others will have drawn unequal spaces between each unit or columns that drift across categorical boundaries.

Goals
- To understand and interpret a simple graph
- To identify and become aware of the need for using equal sized units without overlapping or leaving space between them when organizing data in a graph
- To recognize the need to start at the base line, or the zero point

Materials
- *Berry Picking* storybook
- Data collection from Activity 2, Parts 1 and 2
- Student journals
- Butcher paper
- Plain copy paper, one sheet per student
- Transparency, Mary's Graph: Help Her Fix It
- Transparency, Pete's Graph: Help Him Fix It
- Worksheet, Mary's Graph: Help Her Fix It, one per student
- Worksheet, Pete's Graph: Help Him Fix It, one per student

Duration

Two class periods

Vocabulary

Columns—vertical data or lines

Graph—a display of information through pictures or symbols on a coordinate system

Rows—horizontal data or lines

Preparation

Choose a few student journals that depict common errors in the categorical taste graphs, such as unequal units and spacing. Recreate these by drawing them on butcher paper, or by photocopying the journals and cutting out the graphs. These graphs will be the basis for these lessons, as students continue to modify graphs to make them easier to read and interpret. Make student copies of the worksheets, Mary's Graph and Pete's Graph. You will also need a transparency of Mary's Graph and Pete's Graph. Students will also need a plain sheet of copy paper to complete their homework assignment.

Instructions

1. Read and discuss the *Berry Picking* storybook pages 19–23. Have you ever seen colorful berries?

2. Continue recording daily calendar and weather data in student journals.

3. Review yesterday's work by having students share and explain their journal entries with a partner or group. Use the errors that students made when drawing graphs in their journals as the starting point for this activity.

4. Gather the students around you. Show the recreated graphs from student journals (see Preparation above). Ask, "What can we do to make the graphs easier to understand?"

5. Have students explain what they would change or add to the graph. Ask students to make the changes and manipulate the data by redrawing the graph or reorganizing the cut-out pieces.

6. Use the rebuilt graph to ask students what information they can learn by reading the graph. Then ask, "What would make it easier for you to tell how many students chose each taste?" The goal is to have the

Students Meaningfully Organize Data

Activity 3: Building and Labeling a Graph

students discover that placing numbers on the graph helps them to read the graph more easily.

7. Continue to develop the graph by asking the students what else the graph needs so that we can easily "read it." How do we know what the columns represent? Help the students label and number the graph. See Fig. 3.1 for an example of students making their graph more readable.

8. Tell them that there is one more thing the graph needs. Ask for responses. If they need more direction, ask, "What is this graph about? What should we call it? Can we give it a title?" Discuss titles and how they help in understanding what the graph represents. Have students choose a title for the graph, as they would for a story.

9. Using the overhead projector, display the transparency of Mary's Graph: Help Her Fix It. Begin a discussion about the things that need to be changed in order to "build" this graph correctly. Students may volunteer to make these changes.

10. Hand out the worksheet Mary's Graph for students to use as practice. Have students work in pairs or groups to reconstruct this graph to make it easier to understand, using the transparency for clues. Have each partner recreate this graph in their journal.

11. In different groups, have students share their reconstructed graphs. Students will need to decide which partner will glue this into his or her journal.

12. For additional practice, hand out the worksheet of Pete's Graph. Explain to students that this will be their homework today. Show the transparency of the worksheet for Pete's Graph and allow ample time to discuss the changes that will need to be made in order to correct this graph. As homework, have students draw the rearranged and corrected graph on a blank piece of copy paper.

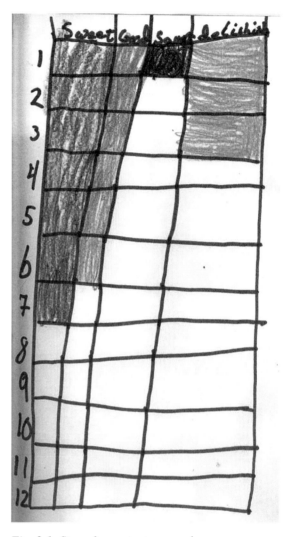

Fig. 3.1: Strawberry tastes graph

Teacher Note

Students may include different starting points for each column, unevenly spaced units, "leaning tower" type columns, and no labels or titles. Remind students that one purpose of a graph is to share information with others. What can they do to display data more clearly for a visitor who is interested in their work?

Teacher Note

Figure 3.2 is a sample graph from a student with partial understanding of the relationship between the number of students who like a particular color (red, blue, or pink) and how to represent it. On the x-axis the student correctly places the value for each color: 8, 5, and 2, but when the student represents it on the graph each color is placed within the same column. It is still possible to make sense out of the student's representation, but with difficulty. This is an example of a student who is still in the process of learning how to correctly graph data.

The form used to support the student's graph creates additional ambiguities. In Figure 3.2 values along the Y-axis are placed mid-way between the column lines. This can cause confusion concerning the students' developing understanding of what is a unit. For example, the student could count each number vs. the space that the number is representing. The recreated form to the right (Figure 3.3) places 'zero" at the bottom left-hand corner (the intersection of the Y and X axis). Units and the numbers representing the unit are aligned. Many students have difficulty understanding a unit and how to represent it (numerically and pictorially) so that it accurately reflects the data.

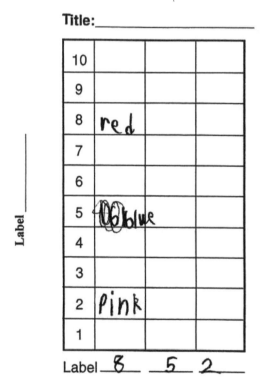

Fig. 3.2: A student's colors graph

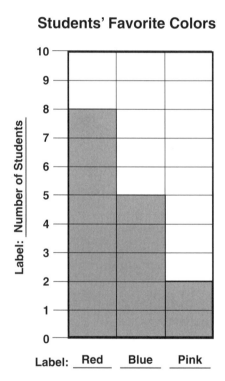

Fig. 3.3: Favorite colors graphs

Mary's Graph: Help Her Fix It

9	5	5
Good	Yuck	Sweet
Good	Yuck	Sweet
Good	Yuck	Sweet
Good	Yuck	Sweet
Good	Yuck	Sweet
Good		
Good		
Good		
Good		

Blackline Master

Pete's Berry Taste Graph: Help Pete Fix It

	Sweet			
Yuck				
Yuck	Sweet			
Yuck	Mushy	Mushy	Mushy	Mushy
Yuck	Sweet			
Yuck				
Yuck	Sweet			
Yuck				

Blackline Master

Activity 4:
Working With Graphs and Tables

In this activity, students refine their understanding of basic measuring and graphing concepts: starting points, same-sized units, equal spacing between units, labeling, and the alignment of categories. Students also explore the concept of symbolic representation by encoding concrete data using symbols that allow for a more portable, smaller graph that can fit on journal paper.

In Activity 4, you provide students with another berry to taste and describe. Students organize their taste responses into categories and sort themselves by their taste words onto a physical graph (a masking tape grid on the classroom floor). Students' organization on the physical graph provides another opportunity for assessment of their conceptual understanding. They may skip a row or not begin at the base line, starting different rows at different points.

Organizing data challenges the students' understanding of what a graph represents. Challenge students to reconsider their organization by asking questions about how to apply numbers to their graphic representations. Through this exploration, students begin to connect graphing to number sense.

Next, students physically arrange themselves into a partially completed table also made with masking tape on the classroom floor. As students find the appropriate location for the taste words they themselves represent, they write their word on a piece of construction paper and leave it, as a reminder, in their place. These reminders allow students to translate the physical table into their journals. As students recreate the table and graph they are faced again with the challenge of scaling (or keeping the data proportional). By making a connection between tables and graphs, students understand that tables both summarize graphs and contain the necessary information to draw one. Later activities further enhance students' understanding of the purposes and inner workings of tables and graphs.

Goals
- To build a graph physically
- To build a table physically
- To connect information between tables and graphs
- To reinforce the need for same-sized units, zero point, and a common starting point
- To organize data by categories

- To evaluate the accuracy of the placement of the data in the physical graph and table and reconcile discrepancies
- To pose and answer questions based on the data contained in a table and/or a graph

Materials

- Berries—one per student
- *Berry Picking* storybook
- Butcher paper
- Large construction paper
- Butcher paper from Activity 3
- Completed homework sheets of Pete's Graph
- Masking tape (for making floor grids)
- Math Tool Kits
- Student journals
- Camera (optional)

Duration

Two class periods

Vocabulary

Starting point—the point from which measuring begins; this is often referred to as the zero point on a ruler

Preparation

Display the modified berry taste graph from Activity 3 in front of the room. Have new berries available for students to taste. Make a 5 x 5 grid on the floor with masking tape (see Figure 4.1); each region should be large enough for a student to stand in. Have large construction paper available. This paper will become labels for the different taste columns in the floor grid. Place a sheet of paper at each row and column of the grid. Also, make a 2 x 5 table on the floor with butcher paper underneath it (see Figure 4.2).

Instructions

1. Read the *Berry Picking* storybook pages 25–29. Discuss the story.

2. Continue recording daily calendar and weather data into student journals.

Activity 4: Working with Graphs and Tables

3. Ask a few student volunteers to share their homework from Activity 3. Ask students to discuss mistakes they found in the Pete's Graph worksheet and ways they fixed those mistakes. Continue this discussion until some of the following topics have been explored: starting point, same-sized units, equal spacing between units, alignment of categories, labeling, and numbers.

4. Allow time for the students to make final corrections to their homework, and have them glue it into their journals.

5. Let the students know that they will taste a different berry today.

6. Hand out the berries, one per student. Students will need stick-it notes from their Math Tool Kits.

7. Have them taste the berry and describe the taste in one word on the stick-it notes.

8. Ask for student volunteers to state their berry taste word. Have volunteers come to the 5 x 5 grid (grid size can vary depending on the number of students in your class) and ask them to write the label (the taste word) for that column on the construction paper and then stand in their place on the grid. Students can skip spaces or start at position 3. Ask the students which category has the most? Is it fair? Discuss.

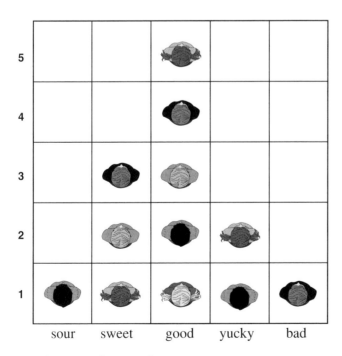

Fig. 4.1: 5 x 5 floor grid

9. Continue to ask for volunteers who have different taste words. When all the different taste words have been stated, have the rest of the students find their places in the grid to create a physical graph (see Figure 4.1).

10. If students run out of columns for taste words, have them solve this problem. They may want to merge categories by combining similar words. For example, students might combine "yucky" and "sour" into a "bad" category. Have the students redefine the category and relabel it on the floor, or they may choose to add a new column.

11. Check to make sure that each student (one student per space) is in the correct column. If possible, take a picture of the graph, which students can then compare to their journal entries and to future versions of this graph.

12. Ask: "What does this graph tell us? How many students thought the berry tasted 'sweet?'" Encourage students to ask questions and other students to answer them. In the process, students will read and interpret the data.

13. Ask students if there is anything else they can do to make the graph easier for others to read.

14. Write the taste words that the students used in their graph such as "yucky," "sweet," etc., as category labels under the "Taste Word" heading on the physical table (see Figure 4.2). Students mark their spots on the table by drawing an "X" and writing their name on the butcher paper where they're standing.

15. Ask them how they can record this information in their journals. This discussion should help them begin the process of connecting graphs and tables from physical representations to abstract graphs.

Taste Words	Number of Students

Fig. 4.2: Taste table

> **Teacher Note**
>
> Frank Hendrickson, a second-grade teacher in Venetie, Alaska, taught his students to discover, for the first time, what a table is and how it works. Hendrickson notes, "They didn't really know what a table was. They knew it had boxes, but they didn't know how to arrange it. One student drew an initial and tally mark for each category. Another used the boxes and wrote his words and tally marks. Another girl beat them both to those ideas and wrote it all with words." The students struggled with this activity, but they better understood the function of a table when they were done.

Activity 4: Working with Graphs and Tables

16. Have the students recreate today's table and graph in their journals. Again, for assessment, notice if students use consistent units, equal spacing, a base line, and labels.

17. Encourage students to discuss how the table relates to the berry taste graph.

Branching Out: Categorizing

Becky Adams, a second-grade teacher from North Pole, Alaska, handed out cards with words relating to a theme (for example, different types of leaves) and had the students decide on three to five organizational categories. Students then formed a graph.

Activity 5:
Building Graphs and Matching Tables

This activity builds on yesterday's work connecting tables and graphs.

Students examine graphs and determine which graph belongs to which table. This exercise will also provide assessment information for students' understanding of the relationship between graphs and tables. After students perform the tasks described in this activity successfully, they are challenged to create a table from a graph. Ultimately, students will see the connection between the abstracted data that a table contains and the phenomena from which it was derived, allowing them to make meaningful interpretations and draw informed conclusions from data.

In Activity 5, students taste another berry and put their taste words on stick-it notes. These words are placed on butcher paper in a categorical graph that the students organize in small groups. Students solve a word problem and then read aloud and complete a worksheet.

Goals
- To create a table and matching graph from oral information
- To reinforce the need for accuracy when transferring data
- To use a base line and iterated units

Materials
- Berries—one per student (optional)
- *Berry Picking* storybook
- Blank sheets of paper for each student
- Butcher paper
- Math Tool Kits
- Student journals
- Butcher paper taste table from Activity 4
- Worksheet, Which Graph is Which, one per student

Start Exploration A: Droopy Plants now

Duration

One class period

Preparation

Display the taste table from Activity 4. Each student will need a copy of the worksheet Which Graph is Which.

Activity 5: Building Graphs and Matching Tables

It also may be time for your class to gather plants, or to plant seeds. This will prepare your class for the lessons in Activity 9, Droopy Plants.

Instructions

1. Read and discuss the *Berry Picking* storybook pages 31–37. Discuss how Panik reacted to the mosquitoes. How did the mosquitoes react to Panik's "quit bothering me" comment?

2. Continue to record calendar and weather data into student journals.

3. Use students' journal entries from Activity 4 of physical (discrete) data. This is a way to review the connection between table and graphs. Ask for student volunteers to explain how the taste table told them what information had to be in the graph.

4. If students do not need more practice, skip to Step 8. If they do, hand out a new berry for students to taste. Organize the students into small groups. Hand out Math Tool Kits.

5. Ask the groups to write a word that describes the taste on stick-it notes.

6. Have the students discuss how to organize their group's descriptive words.

7. Ask the class to provide strategies for organizing these words into a categorical graph. Have student volunteers reorganize the stick-it notes into a graph.

8. Encourage the class to include values (numbers), labels, and a title.

9. Give each student a copy of the worksheet, Which Graph is Which? This worksheet displays two different graphs and one table. Students will need to identify which graph matches the table and draw an arrow between them.

10. Discuss their choice of which graph matches the table. Next, students will need to create a new table that accurately represents the information of the other graph. They can do this on the bottom of the worksheet or on a blank piece of paper. Check and assess students' work. The students will glue this sheet into their journals.

11. Hand out a blank sheet of paper to each student. Read the following new data: "A class in another school did the berry tasting with strawberries. Six students in that class said the berry was yucky. Nine students said it was good. Five others said their berry was sweet." Now students will need to create a table and a graph representing this information.

Teacher Note

Below are two examples of how students choose to make their tables. In Figure 5.1 the student shows a good coordination between data displayed in a table and in a (matching) graph. Although the numerical values are correct, the unit sizes in the graph are uneven. Also, the labeling of the numeric values is not consistently placed. This student is still developing their understanding of units and zero point. In Figure 5.2 there is a developing sense of a table, even though the number display does not match the number of spaces used.

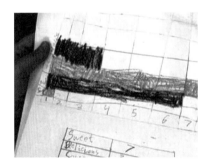

Fig. 5.1: Student's graph and table

Fig. 5.2: Student's table

12. Allow students to choose to work individually, in pairs, or in small groups. Gather and compare their tables and graphs. Have the class offer changes and explanations about how to make these more accurate.

13. Have students glue the new table and matching graph into their journals.

Which Graph is Which?

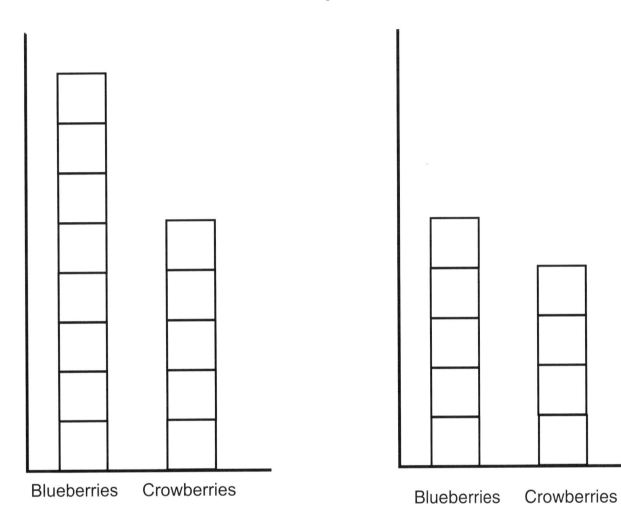

Blueberries	8
Crowberries	5

Blueberries	
Crowberries	

Number of students who like
blueberries or crowberries

Blackline Master

Activity 6:
Describing and Graphing Berry Data

Cultural Note

Linda Brown, from Ekwok, Alaska, reminds us that there are two berry cycles: first, the seasonal order of the different berry types and second, the individual berry cycles—from becoming ripe to overly ripe. When berries are ready for picking depends on the intended use. For example, jam, jellies, and pies will be made from berries picked first because they are full of pectin. Pectin makes berries firm and usually has to be bought from stores to make jam. Later in their cycle, berries are softer, sweeter, and more appropriate for Eskimo ice cream (*akutaq*).

Today's activities build on the work that the students have been doing with discrete data, that is, their work organizing and graphing berry taste words.

Activity 6 challenges students to work with more complex data. Because the data set in this activity includes multiple berries and multiple types of description, students have more responsibility for organizing their descriptive words into a more complex table and graph, comparing the different berries to different senses. An important benchmark for students' developing competence in working with data is their ability to formulate questions related to the data and answer them using the available data. Such questions can be part of a student-generated exploration.

In Activity 6, students describe different types of berries, again on stick-it notes, and categorize them in a table according to attributes of color, taste, and smell. After completing a worksheet matching graphs to tables, they create a graph based on their descriptive tables. Students label their tables and graphs and indicate the values associated with the categories.

Goals
- To categorize descriptive words in a table
- To pose questions that can be answered from the data contained in a table
- To create a meaningful graph based on the data contained in a table

Materials
- Berries:
  Blueberry (or any other available berry)—one per student
  Raspberry (or any other available berry)—one per student
- *Berry Picking* storybook
- Math Tool Kits
- Poster, Berries
- Student journals
- Transparencies, Berries (located in Activity 1):
  Blueberries
  Low Bush Cranberries
  Crowberries
  Salmonberries

Creating Categories

Activity 6: Describing and Graphing Berry Data 61

- Transparency, Berry Description Table (optional)
- Worksheet, Berry Description Table, one per student
- Yup'ik glossary CD [Teacher Reference]

Duration

Two class periods

Vocabulary

Bar graph—graph that uses separate bars (rectangles) of different heights (lengths) to show and compare data

Base line—line on the graph where all data begins

Preparation

Have available two to five different types of berries or other readily available food items that the students can describe. Post the Berries Poster. Prepare the transparencies and the overhead projector. Make student copies of Berry Description Table. Also, consider your needs for Activity 9, Droopy Plants. The Yup'ik Glossary (CD) has information about berry picking.

In Yupiaq stories and other American Indian/Alaska Native stories it is not uncommon for animals to turn into people or for people to turn into animals. For example, in today's reading in the *Berry Picking* storybook there are a few scenes in which mosquitoes turn into people. This can be scary for young readers. The mosquito story needs to be explained to the students as a story within a story. Use your judgment if this part of the story should be skipped, but note that it is presented in its original form so that the story's integrity would be maintained.

Instructions

1. Show poster and transparencies of various berries.

2. Continue to record daily calendar and weather data into student journals. At this point, your class should have collected weather data for 6 days.

3. Read and discuss the *Berry Picking* storybook pages 39–45. What does it mean that the "hunters" were hands? Discuss the story.

4. Pass out a blueberry (or any other available berry) to each student. Encourage students to "study" their berries and to think of one descriptive word, not limited to taste words. Each student will write his or her

word on a stick-it note and hand it to you. Post the notes on the board in an unorganized way.

5. Ask, "How can we organize these different descriptive words?" Allow the students to invent their own ways of categorizing (by color, taste, appearance, texture, etc.). Encourage students to explain their way of categorizing. If necessary, model one category such as color (blue) or texture (smooth). Encourage student discussions.

6. Have students return to their desks and work in pairs. Hand out the Berry Description Table and ask them to take out their journals. Also, post a Berry Description Table at the front board, or use a transparency. Place the students' categories onto the displayed worksheet. The whole class can volunteer, one at a time, to complete the first row of the table posted on the board or on the transparency.

7. Have the students continue to fill out the rest of the table and share their ideas at each step of the process. Ask for volunteers to fill in the blanks on the table at the board or on the transparency.

8. Pass out another type of berry (depending on availability), one to each student. Allow students to work with partners to complete the next row on the Berry Description Table.

9. Next, students may work in pairs or in small groups to complete the Berry Description Table. Students may want to look back to the beginning of their journals when the class first described a berry. Students may discover that their journals are a "memory album" as well as a math journal.

10. Have students formulate questions from the data they have in their tables. What categories did they use? What descriptive words? How many students described the berry as tasting "good"?

11. Ask students to glue their Berry Description Tables into their journals and write their conjectures and answers. Figure 6.1 shows one way to complete the table.

12. For today's journal entry, have each student respond to the following question: "Make believe that Bob, a new student, joins your class tomorrow. I have asked you to help Bob make a graph. What does Bob need to know to make his graph?"

Berry	looks	feels	smells	tastes
salmonberry	funny	cold	juicy	yummy
blueberry	rolly	soft	assirtuq (good)	assirtuq (good)
crowberry	tungupak (black)	hard	lepaituq (no smell)	assirtuq (good)
cranberry				
strawberry				

Fig. 6.1: Berry description table in both Yup'ik and English

Berry Description Table

Berry				
salmonberry				
blueberry				
crowberry				
cranberry				
strawberry				

Cultural Note: Indicators of the Seasonal Berry Supply

Anticipation of berry picking starts months before the immediate preparation for the excursion, as expressed by Margaret Alakayak, an elder from Manokotak:

In the spring, we watch for plants that look like cloudberries (often mistaken for salmonberries). They are called atsaruat ('pretend berries') and are also known as false chamomile or pineapple weed. Atsaruat or 'pretend berries' are used for medicine. When we see those plants in the early spring, we know the berries will be coming.

Margaret adds that the taste of *atsaruat* is also an indicator of berry taste.

A few elders mentioned that 'beargrass,' also called 'cotton balls' (*qivyunguat*), which grows on the tundra, is an indicator of the berry supply (see Figure 6.2). When the fluffy tops show, the berries are ready!

Fig. 6.2: Bear grass

Taking Weather into Consideration

Weather and wind factors affect the seasonal supply of berries. Wind speed on southwest Alaska's tundra can reach 50, 60, or 70 miles per hour, destroying the berry plants as it whips through. Helen Toyukak, an elder from Manokotak, tells us that if the temperature fluctuates too much in the fall, the berries will likely be scarce the following summer. Even worse, she continues, "if the roots of the berries freeze, there will be no berries the following year." Lack of snow cover, which acts as insulation, can leave the roots exposed to cold temperatures.

When there are too few berries to supply the people for the year ahead, alternative arrangements must be made to ensure that the total food supply is not in jeopardy. Helen remembers that in earlier years they would simply pick more greens such as 'sourdock' (*quagcit*) [Rumer arcticus] and 'fiddleheads' (*cetuguar(aq)*) [Dryoteris austriaca] (see Figure 6.3).

People would also dig up mouse burrows to get "mouse food"—tiny berries and roots that the mice had gathered. After the mouse food is cleaned and cooked, it is used to make 'Eskimo ice cream' (*akutaq*) or just eaten plain. All the elders confirm that they always replace the mouse food with other food offerings, and that they never take all of the food from the mice.

Today, mouse food is still used as an ingredient in *akutaq*, even if there are enough berries. Nowadays, says Helen, people who have the money can fly to another area where there are berries to pick.

Fig. 6.3: Cooking fiddlehead ferns

Activity 6: Describing and Graphing Berry Data

Factors Indicating the Quality of Berry Supplies

- Temperature—If temperature fluctuates too much in the fall, berries will likely be scarce the following summer.
- Snowfall—A thick blanket of snow acts as a layer of insulation for the ground, so it protects the hardy plants of the tundra from the wind and from cold temperatures. Lack of snow can leave the roots exposed. If the roots of the berries freeze, there will be no berries the following year.
- Rainfall—Insufficient rain results in berries that are not as plump and juicy.
- Wind—High winds at speeds of 50, 60, or 70 miles per hour can destroy berry plants. In early spring, they can blow the berry blossoms away. In winter, the wind can expose the roots.
- Indications from other plants:
  - *Atsaruat* ('pretend berries,' false chamomile, or pineapple weed)—When these plants appear in the early spring, the berries follow. The taste of *atsaruat* is an indicator of berry taste.
  - *Qivyunguat* or *melquruat* ('beargrass,' cotton grass)—An abundance of beargrass flowers is associated with an abundance of berries. When the cotton grass seeds show, the berries are ready.

Activity 7:
Using Thermometers

In the following activities, students will examine the temperature data they have been collecting. They will learn about various factors besides temperature that may influence the berry supply. Students will determine which recorded temperatures are above freezing and which are below. They will also make predictions about the berry supply based on the basic shape of the temperature graph, thus connecting the symbolic representation with the daily temperature (plotted points). In previous activities students have worked only with categorical or discrete graphs. Activity 7 challenges them to work with continuous data. This data changes over time.

Categorical graphs based on discrete data, represented by the previous taste word graphs, are only concerned with the number of people who chose each category. For example, five students described the berries as sour; sour is the category and five is the number of choices. Continuous data, on the other hand, change continuously from one time to another. Seismic data, for example, are continuous and are continuously recorded. Temperature data also vary continuously but are recorded periodically. These data can be represented by line graphs that show the temperature values recorded at specific times. The line segments connecting the data points represent the general trend of temperature change between the two times, rather than the exact temperatures between the readings.

Some students may wonder about the differences between measuring with rulers and thermometers. In many ways, the two types of measurements are equivalent. In each case you need to define a reference point ("zero") and a unit size. Zero degrees Fahrenheit and zero degrees Celsius are not the same temperature. For example, a student reading a thermometer at 72° just needs to focus on the end point, the line at the top of the liquid, whereas a student reading a ruler at 11 inches must not only pay attention to the end point but also to the starting point. Both the unit size and the zero value are conventions, just as the degree and foot are conventions.

Part 1: Temperature and Thermometers

The first task concerns the partitioning of a paper thermometer. Students choose a meaningful increment and repeat those increments, maintaining their consistent size without gaps or overlapping. They make a mark between every 10 degrees on their thermometers. This task should be met with success for the students. However, asking them to record a temperature that falls between 10 and 20 degrees challenges students to find this location within the 10-degree partitions. Some students may round up or down to

Activity 7: Using Thermometers

the nearest 10 degrees; others may insert another mark midway between the 10-degree intervals. Further, students can examine the actual thermometers they are using to notice that there are four hash marks between each 10-degree interval. What do those marks indicate? Students then decide how accurate their paper thermometers need to be.

Mathematically, students continue to work on partitioning and representation. They will understand that more partitioning leads to more accurate information. This exercise helps students understand partial units, which prepares students for understanding fractional units or fractions.

Students are also introduced to the concept of the freezing point. Relating different temperatures to the freezing point can lead to other arithmetic problems: Is a temperature value more or less than freezing? How much more or less? These kinds of exercises make the connections between graphing and spatial and number sense more explicit.

Individual or small group work provides an opportunity for assessment of students' conceptual understanding of units. This activity further develops the connections between units, iteration of units to accumulate distance (value/temperature), spatial reasoning, and basic number sense.

You should remind students that the data they are working with in this activity—temperature—will help them understand more about plant growth and berries in particular.

Goals
- To practice representing data as a line graph
- To become aware that temperature affects berry production
- To construct paper thermometers in different increments and resolve issues related to partial units

Materials
- *Berry Picking* storybook
- Math Tool Kits
- Student journals
- Construction paper, one sheet per student
- Transparency, Thermometer with Fahrenheit Increments of Ten Degrees
- Transparency, Thermometer with Fahrenheit Increments of Two Degrees
- Worksheet, Thermometer with Fahrenheit Increments of Ten Degrees
- Worksheet, Thermometer with Fahrenheit Increments of Two Degrees

Units of Measure

Teacher Note

The formula for converting Fahrenheit to Celsius is:

(Fahrenheit − 32) x ⁵⁄₉

while the formula for converting Fahrenheit to Celcius is:

(Celsius x ⁹⁄₅) + 32

For example, take the temperature 45 degrees Fahrenheit, then subtract 32 degrees and multiply the answer by ⁵⁄₉:

(45 − 32) x ⁵⁄₉ = 13 x ⁵⁄₉ = ⁶⁵⁄₉ = 7 ²⁄₉ C

Conversely, take 20° C and convert it into Fahrenheit:

20 x ⁵⁄₉ = 36 + 32 = 68 F

Duration
One class period

Vocabulary

Thermometer—an instrument for measuring temperature, especially one having graduated measures along a glass tube filled with a liquid that expands as temperatures rise

Degree—a unit used when measuring temperature

Preparation

You will need transparencies for the Thermometer with Fahrenheit Increment of Ten Degrees and the Thermometer with Fahrenheit Increments of Two Degrees. The students will also each need a copy.

Also, consider your needs regarding the plants to be used in Activity 9, Droopy Plants.

Instructions

1. Continue recording the daily calendar and weather data in student journals.

2. Inform students that they will study the thermometer and learn how to read temperatures above and below the freezing mark. Remind students that cranberries and crowberries can be picked even after freezing temperatures begin; sometimes people pick them in the spring after a long winter.

3. Gather students around the overhead and display the transparency, Thermometer with Fahrenheit Increments of Ten Degrees. You may choose to use either system, Celsius or Fahrenheit, by simply covering over one side of the thermometer transparency with a sheet of paper. The Celsius system may be easier. However, if your students are more familiar with Fahrenheit you may choose to start with this system. Ask the students what thermometers measure. What is this thermometer missing? Ask, "What do the lines mean? What does the red liquid do?" Discuss measuring temperatures and degrees.

4. Ask a student volunteer to label the unmarked lines on the thermometer shown on the overhead. Note that on this graphic the lines on the Fahrenheit side move by tens, while the Celsius side moves by fives. Ask another student to draw in the liquid for a temperature the class chooses. Repeat this process several times, drawing and reading several temperatures using units of 10 (10, 20, 30, etc.).

Activity 7: Using Thermometers 69

5. Hand out the worksheet, Thermometer With Fahrenheit Increments of Ten Degrees to each student, or have students draw their own blank thermometers (see Figure 7.1). Ask the students to draw in a temperature for 55 degrees Fahrenheit.

6. Ask the students to determine, in small groups, how many partitions, or divisions, there should be between each 10 degrees. Ask each student to draw in the lines and explain the values for the number of lines they choose.

7. Project on the overhead the Thermometer With Fahrenheit Increments of Two Degrees transparency. Focus attention to the lines between 60 degrees and 80 degrees on the Fahrenheit side of the thermometer. Ask, "Why are these lines there? What do they mean? Do they have a value? What is the missing number at the bold line between 60 and 80? How would we show 65 degrees?"

8. Ask students why it may be important for us to be able to tell the temperature as accurately as possible. Pass out the matching worksheet Thermometer with Fahrenheit Increments of Two Degrees. Have students work independently to correctly number the degree lines. Students then return to the whole group and explain their reasoning. Allow time for some students to correct their work and glue their sheets into their journals. Have students continue to work in groups, help each other, and explain to each other how many increments they will need to accurately and easily read their "thermometer." See Math Note on the next page.

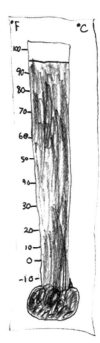

Fig. 7.1: Thermometer drawing

Teacher Note

Organize students in collaborative groups to resolve mathematical issues related to problems that they will encounter as they partition units. (See Math Note on next page.) You may want to review counting by twos, fives, and tens since these are the increments used for the thermometers created for these activities.

Math Note

A few teachers reported that students used different strategies to partition the spaces between 50 and 60 degrees. They knew that the difference between 50 and 60 degrees is 10. Further, some said that 5 and 5 is 10, therefore there need to be 10 spaces between 50 and 60.

These different understandings create a good teaching opportunity. One way to help students reconcile the difference between the representations is by having one student count lines up from 50 by ones until they reach 5 and mark that place. Have another student count down from 60 by ones until they count 5 spaces and mark that place. This will perplex the students if they also choose to count lines instead of spaces. Similarly, Figure 7.2 shows two different representations, one correct and one incorrect. Have the students discuss these two representations and resolve the differences between them.

A second reason some students get into this predicament is because they confuse the starting point on a ruler or thermometer. For example, does the count for the measure begin in the space or at the hash mark? These are a few of the common errors students make. Have the students reconcile this in small groups and have them discuss and resolve these misunderstandings.

After your students succeed in partitioning the thermometer to show 55 degrees F and you review the way they partitioned their thermometer, you can ask if any students tried to partition their thermometer in increments of two degrees and find 56 degrees F. Students may have difficulty reconciling 10 and 2 degree increments on their paper thermometers. This presents an interesting challenge to students as they decide on their strategy for partitioning their thermometer.

Also, keep in mind the purpose for measuring and that the degree of accuracy (measuring as approximation) relates to how many increments are sufficient for the purpose at hand. In other words, if accuracy within 5 degrees is sufficient, then it would not be necessary to partition to very small increments.

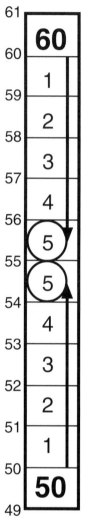

Fig. 7.2: Two different partitions of a thermometer

Activity 7: Using Thermometers

Part 2: Temperature and Freezing Point

The concept of partial units represents a milestone in this module for students' understanding of measuring as well as foundational knowledge concerning fractions. Students encounter partial units in the following lesson and later on in the module when they measure their height and shadows. Typically students create partial units by folding or dividing units into smaller subunits. They may think of each unit as a whole or as a part unit of the larger unit.

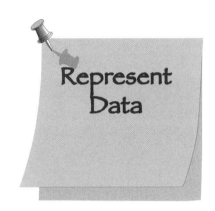

Students will collect temperature data from three work centers with containers of water at different temperatures. Students record these temperatures and incorporate them into a table and a graph. They perform a new task as they create a line graph. They arithmetically compare their collected temperatures to the freezing point, as they had opportunity to do in the first part of this activity.

Goals
- To practice reading, interpreting, and recording temperatures including partial values (units)
- To learn that zero degrees Celsius and 32 degrees Fahrenheit is the freezing point
- To write meaningful questions that can be answered by the data
- To represent temperature data as a line graph

Materials
- Clear cups, one per three water stations
- A small amount of salt for each water station
- Three small thermometers, one per station
- Ice, one cup for each station
- Hot water, one cup for each station
- Warm water, one cup for each station
- Blue crayons
- Thermometer worksheets from Part 1
- Worksheet, Horizontally and Vertically Oriented Temperature Tables, one per student
- Student journals
- Math Tool Kits

Duration
Two class periods

Vocabulary

Freezing Point—the temperature at which water begins to freeze, which is 32 degrees Fahrenheit and 0 degrees Celsius

Preparation

Arrange three or more centers, depending on the size of your class, for water temperature experiments. Prepare one center with a cup filled mostly with ice, and a little salt water; one with hot water; and one with warm water (see Figure 7.2). Label the centers Day 1, Day 2, and Day 3. Make student copies of Thermometer with Fahrenheit Increments of Two Degrees that the students numbered in Part 1 (three for each group of students) and Horizontally and Vertically Oriented Temperature Tables worksheets (one per student).

Instructions

1. Ask students to work in small groups at the water centers (see Figure 7.3). They should use the "thermometers" from Part 1 on which they've drawn degree lines to record temperatures at each water station. Students will use a pencil or crayon to indicate the temperature of each water center on their thermometers.

2. Students will need to visit each water station and record the temperatures they find in each cup on their Thermometer with Fahrenheit Increments of Two Degrees sheets.

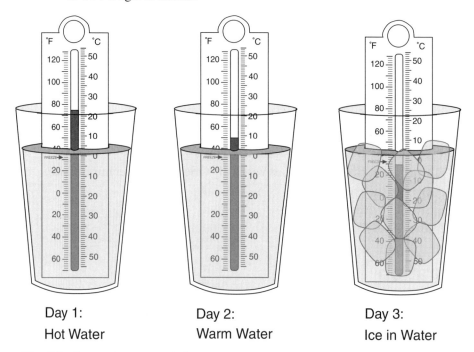

Day 1: Hot Water

Day 2: Warm Water

Day 3: Ice in Water

Fig. 7.3: Cups of water and thermometers for three water stations

Activity 7: Using Thermometers

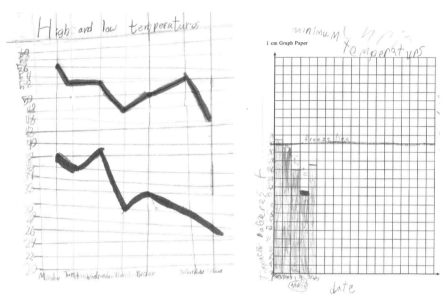

Figs. 7.4 and 7.5: Students' graphs

3. Hand out the worksheet, Horizontally and Vertically Oriented Temperature Tables. Ask if there is a difference between the information represented in the different orientations. Note that students can become accustomed to the "one right way," which becomes an idealized form.

4. Have students transfer the information from their paper thermometers into both vertical and horizontal tables. Make sure that students recognize that the information shown on both tables is the same, and that both are correct.

5. Ask the small groups to create a graph from the information on their tables. Encourage student discussion within the small groups.

6. Allow students to share their paper thermometers, tables, and graphs with the class. Discuss the importance of using equal-sized units and partial units. (Note students' strategies in representing temperature data. Do they round up to the nearest half unit, whole unit, etc. Do students represent the same temperatures differently? If so, have them understand the reasons for the differences and have them reconcile them.)

7. Repeat the information that people harvest cranberries and crowberries even after freezing. Have students use a blue crayon to mark the freezing temperature on their graphs. Ask students to read the freezing point on their thermometers in both Celsius and Fahrenheit.

8. Discuss freezing temperatures and how they relate to the table or graph the students have just created. Ask students to look at their water center data as if it represented outside temperatures and discuss which day (the stations are labeled Days 1–3) the cranberries and crowberries would

begin to freeze and would then be ready to harvest. Relate the students' experience of recording the water temperature to their collected weather data and to berry harvesting.

9. Have students take out their journals and turn to the data they have been collecting during the daily weather and calendar activities. From the data gathered in their Observing the Weather and Collecting Data pages, display on the board the temperatures from any five sequential readings. Demonstrate, using the first day's and second day's temperatures, how to graph this temperature data on a line graph instead of a bar graph. Be aware that students have much less experience with line graphs. Teach students that a point on a graph represents two values: the temperature on the Y-axis and the day or the number on the X-axis. Ask a student volunteer to graph the second day's data (temperature) onto the graph. Draw a line between the points and explain that this can be done because the information is continuous, that is, constantly changing over time.

10. In their journals, have students recreate the started line graph and complete it by inserting the final three days of temperature data and connecting the points with lines. Challenge students to color in the line representing the freezing point. Would cranberries be ready for picking based on the weather data you collected?

11. Ask students to work in small groups to create three questions that can be answered by their graphs. Ask each group to write their questions in their journals. Share the questions with the class and let students answer a few of the questions.

12. Conclude today's lesson to prepare for measuring and continuous data. Gather students as a class to discuss different types of things that can be measured. They know height and length can be measured with feet, inches, or even yardsticks and that temperature can be measured in degrees with a thermometer. Ask them if they can think of types of things that can be measured in different ways. As a fun homework assignment, encourage students to ask family members for their answers and bring them into class—one student returned to class the next day sharing his information on earthquakes and the Richter scale!

Thermometer with Fahrenheit Increments of Ten Degrees

F°   C°

80   30

60   20

     10

40

→   0

20

Blackline Master

Thermometer with Fahrenheit Increments of Two Degrees

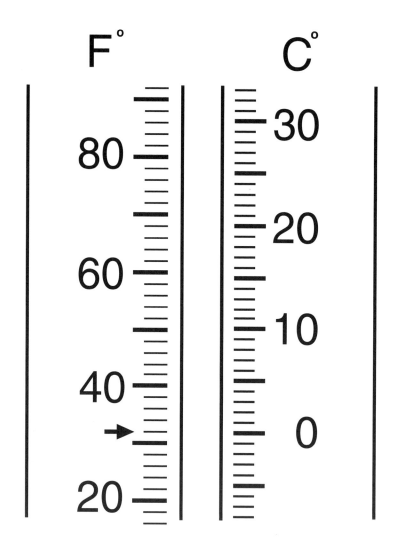

Blackline Master

Horizontally and Vertically Oriented Temperature Tables

Day			
Temperature			

Day	Temperature

Blackline Master

Activity 8: Observing the Weather and Collecting Data

As part of the daily weather and calendar activities, your class has observed and participated in the process of data collection and data recording. However, most of the responsibility for these activities has been with the teacher. In this activity, students become more independent and work with more complicated data. Further, the class applies its developing understanding of the math concepts to some of the factors that affect berry growth.

This lesson challenges students' conceptions of recording and representing data and then interpreting that data. Are they able to understand and perform these tasks with assistance? Are they able to work independently and help others? Students' knowledge varies from not knowing how to record data in a table to using unequal units in a graphic representation. Some students won't understand that units must be linked without gaps or overlaps. Some may not be able to express a concrete reading of the data; for example, from the data set they may not be able to verbalize that on Monday the temperature was 56° Fahrenheit and the sky was cloudy.

One way of assessing students' understanding revolves around whether or not they can pose their own questions and identify strategies to answer them, as well as their ability to evaluate the appropriateness of those questions. As students work with this more sophisticated data set—temperature, precipitation, and sky conditions—they can observe the patterns, trends, and relationships of these phenomena. In addition, this lesson gives you an opportunity to assess students' command over the concepts of numbers and measurement and the relationship of those concepts to graphs and to their daily experience.

In Activity 8, students use the weather data they have been collecting to be able to identify patterns. They create tables to organize their data and graphs to represent their tables. If some students are still unsure about how to organize their data, you should continue to model these activities and use more experienced students to help those less confident with the material.

By anchoring data in exercises that relate to everyday experiences (weather), students can make a personal connection to data. At the end of the module, they will review their data and predict plant growth (more specifically berries) for the upcoming season.

Activity 8: Observing the Weather and Collecting Data 79

Goals

- To identify weather factors influencing berry production
- To create weather data collection forms
- To create graphs from collected weather data
- To be able to organize weather data into tables and graphs and to recognize that the same data can be represented in multiple ways

Materials

- *Berry Picking* storybook
- Worksheet, Weather Table (optional, one per student)
- Student journals
- Observing the Weather and Collecting Data sheets from Activity 2

Duration

One class period

Preparation

In the last lesson, students made a graph for five days of data. Now they will build a table and include additional variables as shown in Figure 8.1. Use data collected as part of the daily weather activities. If your class has reached the developmental point at which they are ready to create their own forms of data representation, encourage them to do so. However, we also provide a table form for this purpose. Remember that the simplest table has only one variable, as in the last lesson. Each variable a student will add creates a new column or row. Have your students use no more than three variables at this point in their development.

Instructions

1. Review *Berry Picking* storybook, page 5–9 (when berries are ready). Read pages 47–51.

2. Have students take the weather data they have been gathering as part of their daily calendar and weather activities. Students will take five days of this information and put it into table form. We encourage students to create their own tables using appropriate labels as in past lessons. However, the Weather Table is provided if needed. Have the students work collaboratively as they create their own tables.

3. As in Activity 7, discuss different orientations of tables. If you feel students are still having difficulty with this concept, have them practice by altering the orientation of the table they have created.

Teacher Note

If your students are ready to create their own weather collection form or table, allow them time to create, share, modify, and finalize the class's form or table. Otherwise, you can create a table for them, similar to the one shown in Figure 8.1.

Day	Temperature	Rainfall	Cloud Cover
1	41	0	Sunny
2	43	2	Partly Cloudy
3	38	5	Overcast
4	33	0	Partly Cloudy
5	29	0	Sunny

Fig. 8.1: Temperature, rainfall, and cloud cover chart

Analyzing Data

Assessment Suggestions

This is a good place to assess students' ability to read, interpret, and predict patterns in tables. Can your students read the table? Do they know what the headings represent? What the numbers represent? If not, then more practice of various kinds is needed.

4. Discuss the information students recorded and its connection to berry production.

5. Students should have previously made temperature graphs from yesterday's lesson. Ask students, in small groups, to represent their weather data tables as graphs. Ask each group to make one graph of precipitation and one of cloud cover over five days. Remind students about labels, titles, unit sizes, zero points, and spacing.

6. Ask the groups to share their tables and matching graphs. Is the information the same in the table and the graph? As a class, discuss the graphs and how to improve them.

7. Ask the class if they notice any patterns in the data (when reading the table, students should be able to state the temperature, precipitation, or cloud cover conditions for any day). Ask students to interpret from the collected data which day was the warmest, coldest, cloudiest, or wettest. They may see relationships between cloud cover and precipitation, or cloud cover and temperatures.

8. Display the groups' work and ask students to record, individually, the three different graphs (rainfall, cloud cover, and temperature) in their journals. They may also want to include the completed weather data collection table in their journals.

Weather Table

Day	Temperature	Rainfall	Cloud Cover
1			
2			
3			
4			
5			

Activity 9:
Droopy Plants

Linda Brown, a Yup'ik teacher from Ekwok, Alaska, says, "Whenever I complained about too much snowfall, my mom would tell me, 'Lots of snow is good because we will have lots of berries in the fall.'"

Throughout this module, your class has been observing and recording weather data. This data allows students to observe trends in temperature and the relationship between temperature and cloud cover and possibly precipitation. In Activity 9, students relate measuring, graphing, and weather observations to plant growth.

This science-oriented activity yields rich data for students to collect, record, and analyze in their journals. The exercise challenges students to determine the amount of water required to keep a household plant healthy and growing and to find out if local precipitation would be sufficient to keep the plant healthy. Students apply their measuring skills when watering their plants and observing changes in the plants' height and health. Students learn that this type of measurement yields data that can be analyzed with tables and graphs.

When students measure plant height they face the problem of starting point, or where they should begin measuring. Plant soil may not be evenly distributed within the pot, so measuring the same plant may yield multiple results. Even if students use their rulers accurately, they will need to solve this problem to get comparable, useful data. During the week of this activity, students use their journals to illustrate and describe changes that take place when plants receive a given amount of water.

Fig. 9.1: A bountiful berry harvest (photo courtesy of Nastasia Wahlberg)

Activity 9: Droopy Plants

Students water their plants according to the weather patterns they observe. They can use plants they bring from home, but seeds planted at the beginning of the module should be ready for use in this activity.

These exercises raise the opportunity for students to make well-grounded conjectures. For example, is the observed precipitation for a day or week enough to keep a plant healthy? Students learn that snowfall and rainfall affect the berry supply. Plenty of water early in the berry plant's cycle yields plump berries. If the rain does not come until the berries start to ripen, it cannot help the berry crop.

Goals
- To make conjectures about plant growth and their need for water
- To conduct an experiment to determine plant needs for water, measure water (volume), and make observations about changes in plant health and/or growth
- To organize, interpret, and draw conclusions from data collected about a plant's need for water

Materials
- *Berry Picking* storybook
- Measuring cups
- Plants, one per group of students
- Student journals

Duration
One class period for preparations, but the experiment continues for five consecutive days.

Vocabulary
Moisture—another word for water, and a term for water density
Precipitation—any form of water or moisture that falls from the sky
Roots—the underground part of a plant that absorbs nutrients
Soil—another word for dirt

Preparation
By now you will have gathered the necessary plants for the experiments of this activity. Each group of students should have its own plant. If no seeds were planted at the beginning of the module, then use plants that students donate from home and/or plants that may be in the classroom. One teacher found strawberry sprouts at a greenhouse and her students transplanted them and eventually ate the berries that they produced. This particular

strategy makes the connection between berry growth and production and precipitation quite direct.

Some teachers may want three or more plants of equal size so their students can perform several experiments to determine the effects of different factors, such as water, light, or heat.

Instructions

1. Read the *Berry Picking* storybook, pages 53–61, and discuss what is the meaning of this story.

2. Ask questions about how weather affects the berry plants: What might happen to the berry patches if they don't have very much snow cover during the winter? Why would the amount of snow make a difference to berry plants that grow in the summer? What happens to berry plants that receive too little rain? What happens to plants that get too much water? If a particular question or conjecture interests students, note it on the board.

3. Gather students around the plants. Announce that although these plants will be growing in the classroom, you want them to pretend they are berry plants growing outside in a berry patch.

4. Ask students what happens when the warm spring sun shines on the snow that covers berry patches in the winter.

5. Announce, "We want to see what happens to berry plants if they do not have enough snow or rainwater." Ask for student input on how this might be accomplished with the plants inside the classroom.

6. There are several ways to approach this classroom experiment. One way would be to begin with a fairly dry, but not wilted, plant. Set a measuring cup outside to collect actual precipitation for watering the plants. A second option would be to refrain from watering the plants until they are quite wilted, but not yet dead. Then ask the class to decide how much water to give the plants to make them become healthy again. Another method would be to ask each group to give their plant a specific amount of water, amounts that would vary with each group. Choose a method that would work best for your classroom situation.

7. This is a good place for students to make conjectures regarding their plant and the method you have chosen to water them. They should determine how they will answer these conjectures. Ask students to think about what will happen to their plants. Students will need to observe what plant changes occur. These observations should include plant

Fig. 9.2: Wilted plant

Teacher Note

As students water their classroom plants, they record the amount and frequency of watering as well as measurements of plant growth (height). By making these observations, students continue to revisit measuring concepts such as units, equal spacing, starting points, and the use of standard measures.

Activity 9: Droopy Plants

height, plant condition, color changes, leaf or wilt changes, or anything else the students notice.

8. Have students partition a page in their journals into sections and label the sections Day 1 through Day 5. Daily, have the students make observations concerning how much water and plant growth and make journal entries on the plants' progress. Also, you may ask them to draw their plants.

9. Have students in each group reconcile differences in measuring (see Teacher Note). Hold a class discussion, comparing the changes in the plants and what may have caused it. Discuss their predictions and if they were accurate. Encourage students to compare plant growth to the data that they have been collecting.

10. Have the students record what they have learned about the relationship between water and plants and the accuracy of their conjectures in their journals.

> **Teacher Note**
>
> When students in the same group measured the height of their plant in John Purcell's second-grade class, they noticed that they had different measurements. Purcell asked them why. The students realized that they were measuring their plants at different starting points. The students measured the height of the plant using different starting points, so they had different values for the height of the plant (see Figure 9.3). Issues of accuracy of measuring and common starting points are a few of the measuring issues that students will continue to explore throughout the module.

Fig. 9.3: Common starting points

> **Teacher Note**
>
> Becky Adams, a second-grade teacher from North Pole, Alaska, asked individual students within each group to take turns being in charge of the plant. Students came to realize that too much water was not good for plants, either, and that plants don't survive rough handling. One student commented, "I wonder how any plants out there ever make it?"

Activity 10:
Where Do Berries Come From?

In previous lessons students turned random data into a meaningfully organized graph. As they created graphs, students learned that units should be of equal size and that columns and rows should be uniform and labeled. In this activity students will learn to read a map. Maps are actually similar to graphs in that they both represent locations with a coordinate system that is based on a grid. Students may not immediately understand this connection; therefore, the following tasks begin to develop this understanding.

Activity 10 introduces your students to the geography of Alaska and continues to explore the Central Yup'ik Eskimo culture of southwest Alaska. Students view various maps and learn to recognize the shape of Alaska as they practice using the four cardinal directions. Students practice locating places and resources on a map.

Part 1: Map Studies

Goals
- To learn geographical facts about Alaska
- To see how Alaska relates geographically to the rest of the United States
- To practice using the directional words of north, south, east, and west
- To consider what types of berries grow locally and also what types of berries people depend on in southwest Alaska

Materials
- *Berry Picking* storybook
- Large pull down map of the U.S. and Alaska (optional)
- Student journals
- Transparency, Alaska
- Transparency, Kuskokwim River
- Transparency, North America
- Transparency, Map of the United States with Alaska Superimposed
- Worksheet, Alaska, one per student
- Worksheet, Map of the United States with Alaska Superimposed, one per student
- Worksheet, North America map, one per student
- Math Tool Kits

Activity 10: Where do Berries Come From?

Duration
Two class periods

Vocabulary
Cardinal Directions—north, south, east, and west
Compass Rose—a symbol on a map that shows direction
Map Key—tells what the symbols on a map stand for

Preparation
In the lesson, you may want to display artifacts, charts, pictures, books and maps relating to Alaska. Keep a world globe on hand and a wall map of the world and/or map of the U.S. pulled down in front of the class. You may extend any of these lessons by drawing and labeling mountain ranges, rivers, or even shading in the areas where the Yup'ik people live. Make student copies of the Alaska map, the map of North America, and the map of the United States with Alaska superimposed.

Instructions

1. Review what has happened in the *Berry Picking* story so far and how to tell if berries are ripe. Review the seasonal indicators of berries in *Berry Picking* storybook or refer to the Cultural Note in Activity 6 (page 64).

2. Ask, "Where did this story take place?" Take some student guesses. Using the overhead projector, display the Kuskokwim River transparency. Remind students that the story originates from actual childhood experiences in this area.

3. Show the transparency of Alaska or use a globe or classroom map of Alaska.

4. Give each student a copy of the Alaska map. Choose a few points of reference for the students, depending on where you are located. You may want to include Juneau and the village or town where you are.

5. Have students name places that should be noted on their maps. Write these on the board. Ask students to draw a dot on their Alaska worksheet to locate these places in Alaska, and write place names next to the dots. You can model with the transparency while students help each other. Discuss the "Panhandle" and the Aleutian Islands as important sections of Alaska (students tend not to see them in connection with the rest of the state).

6. Hand out copies of the North America worksheet. Draw student attention to the transparency of Alaska and have students find Alaska on the new map. Students will once again need to locate and mark the same points of reference that they used on their map of Alaska.

7. Question the location of Alaska in relation to other areas on the map. Questions should be designed to improve map skills, particularly the correct use of cardinal directions. You may discuss the location of Mexico and Canada as well.

8. Show the transparency Map of the United States with Alaska Superimposed. Ask students if they recognize Alaska. Give each student a copy of this map. Have them color in Alaska while discussing the points of reference they located on the previous maps using the cardinal directions of north, south, east, and west.

9. Ask students to glue these completed maps into their journals.

Part 2: Using a Grid Map

The purpose of this section is for students to select data by choosing three resources displayed on the Togiak Resource overlay transparency (for example, one mammal, one bird, and one sea animal) and copy the resource symbols to their individual maps.

Goals
- To identify and classify animals of southwest Alaska
- To learn what a local resource is
- To learn aspects of subsistence living
- To use a coordinate grid system and ordered pairs to locate local resources

Materials
- *Berry Picking* storybook
- Maps, various geographical and political (teacher provides; see Preparation for Part 1)
- Student journals
- Transparency, Togiak Region Map
- Transparency, Togiak Region Map with Grid
- Transparency, Togiak Resources Overlay
- Worksheet, Togiak Region Map, one per student
- Worksheet, Togiak Region Map with Grid, one per student

Activity 10: Where do Berries Come From?

Duration
One class period

Vocabulary
Density—the amount of a substance contained within a specific area

Grid—a picture divided into equally spaced squares

Locating—finding the relative position of an object in space by using a coordinate system or relative position

Natural Resource—plants, animals, and other materials that can be harvested for human use

Ordered Pair—a point on a graph or grid; the point represents the value on both the x- and y-axes

Preparation
Post copies of the Togiak Region Maps with and without the grid and the Togiak Resources overlay in the classroom. Each student will need a copy of the Togiak Region Map with and without the grid. Also, you may want to display various types of maps with different grid and coordinate systems, for example, geographical and political maps in telephone books, topographical maps, USGS maps, city maps, atlases, and globes.

Instructions

1. Read the *Berry Picking* storybook pages 63–71. Kang'aq picked her first berries. How did she and her family celebrate this? Have the class discuss. Note: in some cultures who you are named after is very important. In this example, Kang'aq is treated like her namesake. Ask students who they are named after and does that have special meaning. Because Kang'aq is named after a deceased person whose husband is still alive, Kang'aq is treated as if she was that husband's wife in certain ways. How do Aanaq and Kang'aq treat the elder?

2. Gather the students into a large group and show the transparency Togiak Region Map without the grid. Discuss the map and locate villages, rivers, lakes, islands, etc. (Additional activities here may include drawing a compass rose or creating a key for the map.)

3. Display the Togiak Resource Overlay over the Togiak Region Map without the grid, showing the natural resources of the Togiak Region. This overlay includes the locations of these resources: berries, land mammals, sea mammals, salmon, and birds.

4. Hand out copies of the Togiak Region Map without the grid.

5. Ask students to pick one resource from the overlay. Locate it and then draw it on their Togiak Region Map. (The purpose of this particular section is to have the students understand the difficulties of trying to locate without a grid, and then the benefits and uses of the grid.)

6. Ask students to find partners. One partner will choose a particular resource and attempt to describe its specific location so that the other student can draw it on their own region map. (This will be difficult to do without a grid format.)

7. Ask the students how to explain the location without looking at one another's maps. Allow time for individual discussions and then record the students' ideas on the board.

8. Display the Togiak Resources Overlay over the Togiak Region Map With Grid. Discuss how the grid helps us locate the resources. Ask, "What else do we need to make this grid really usable?"

9. Hand out copies of the Togiak Region Map with Grid to each student. Ask the students to choose another resource and locate it on their maps. Have the students tell their partners where they located their resources without showing their maps. See if their partners can more easily find the locations with the grid.

10. Placing the Togiak Resources overlay over the Togiak Region Map with Grid, have the students individually locate one or more of each resource type on their Togiak Region Map with Grid. The Togiak Resource Mapping Game provides practice in using a grid with the alphanumeric system to locate resources.

> **Teacher Note**
>
> The goal is for the students to realize that a coordinate system helps locate objects on a map. Students may develop their own coordinate system or you may introduce the alphanumeric system.

> **Teacher Note**
>
> Nancy Sharp of Manokotak had her students draw a map on the floor using chalk and make the islands by placing them into the grid. John Purcell, in Fairbanks, had his students use a coordinate system to find the location of places. Others have had their students create a coordinate system, which helps them begin to understand scale.

Activity 10: Where do Berries Come From?

Teacher Note

Maps are often presented in two forms:

1. a grid with numbers on both axes. For example, (2, 3) representing the intersection of 2 on the x axis and 3 on the y axis, or
2. a grid with numbers and letters not placed on the grid lines, but between them. For example, (2, C) indicates a range.

In the latter case order does not matter since (2, C) and (C, 2) are equivalent, while in the former case order does matter, since (3, 2) and (2, 3) indicate different locations. In this activity, students practice locating objects on a grid by finding a range.

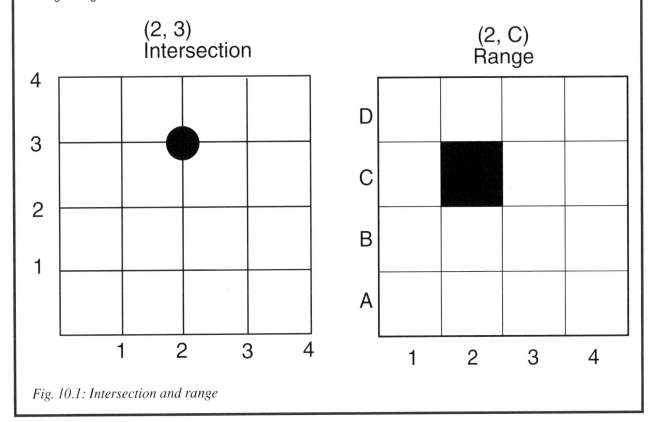

Fig. 10.1: Intersection and range

Part 3: Togiak Resource Mapping Game

Goals
- To use a coordinate grid system to locate local resources

Materials
- Transparency, Togiak Resource Mapping Game
- Worksheet, Togiak Resource Mapping Game, one per group of students
- Handout, Togiak Resource Mapping Game cards, one shuffled pile per group of students

Vocabulary
Range—a grid with numbers and letters along each axis, indicating an area of the map.

Intersection—a grid with numbers along each axis determining a specific point.

Preparation
Copy the Togiak Resource Mapping Game for students to use in small groups. Copy and cut a batch of resource cards. Shuffle them before handing them out to students.

Instructions

1. Project the transparency, Togiak Resource Mapping Game. Highlight the region (3, C).

2. Discuss the coordinates and how they are used. Highlight all of the regions that are in row "C". Then ask for a student volunteer to highlight all the regions found in column "3". Focus attention on where "C" and "3" meet. Explain that this is called "3, C". Ask students to name the resources located in (3, C). Practice this with several other regions on the map.

3. Divide the class into teams of three or four students. Pass out the resource cards, shuffled randomly, to the groups.

4. Teacher calls out a region, such as (4, D). The groups will locate that region on their map and determine if they have a resource card that matches a resource in that region. The group then places the matching card over this region.

Activity 10: Where do Berries Come From?

5. The team that places all of their cards first wins.

6. This same map can also be used for a riddle game. For instance, the teacher may say, "I am in column 5. I am the only resource in my region. What am I? Where am I?" Students or small groups may take over and write or say more riddles for the class.

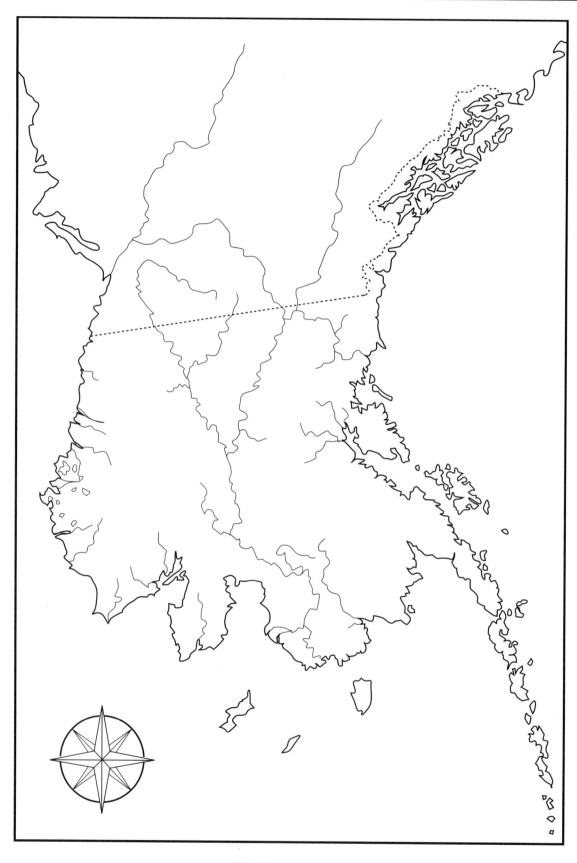

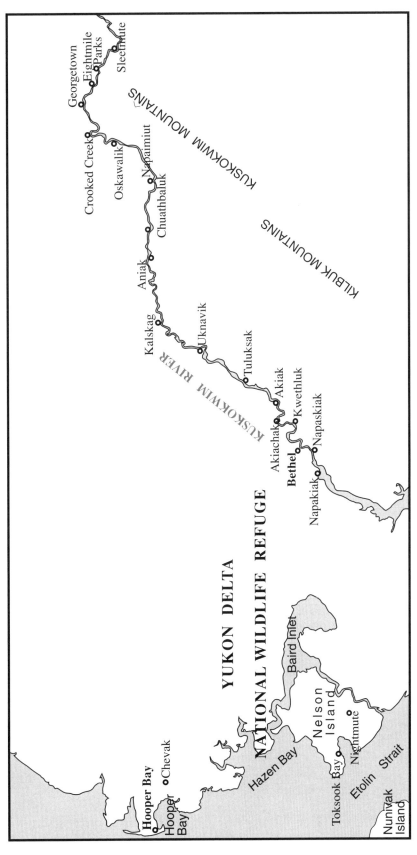
Blackline Master

North America

Blackline Master

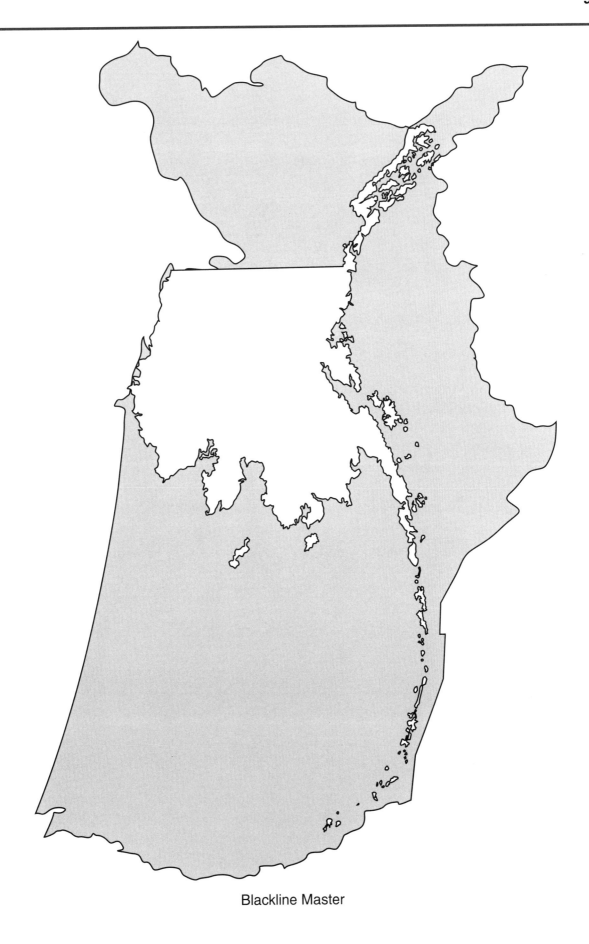

Map of the United States with Alaska Superimposed

Blackline Master

Togiak Region Map

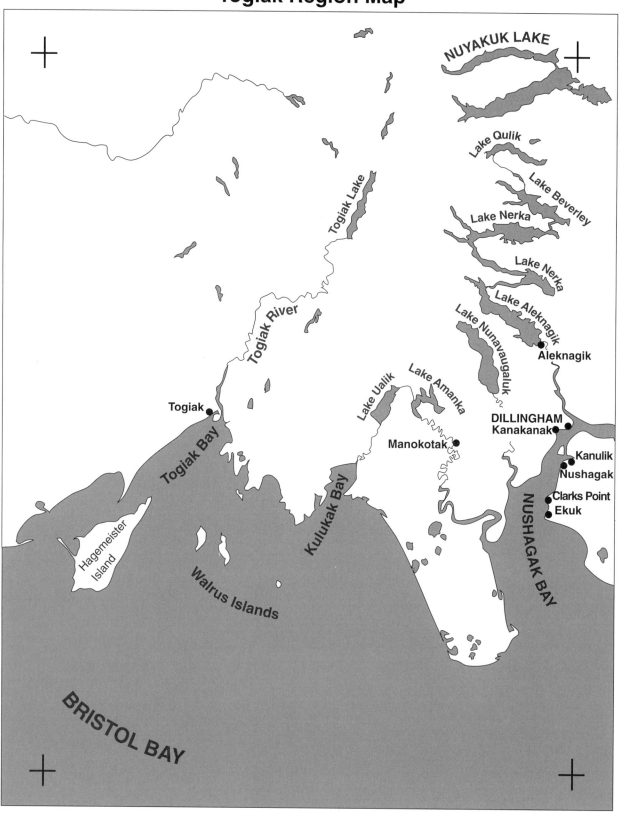

Blackline Master

Togiak Region Map with Grid

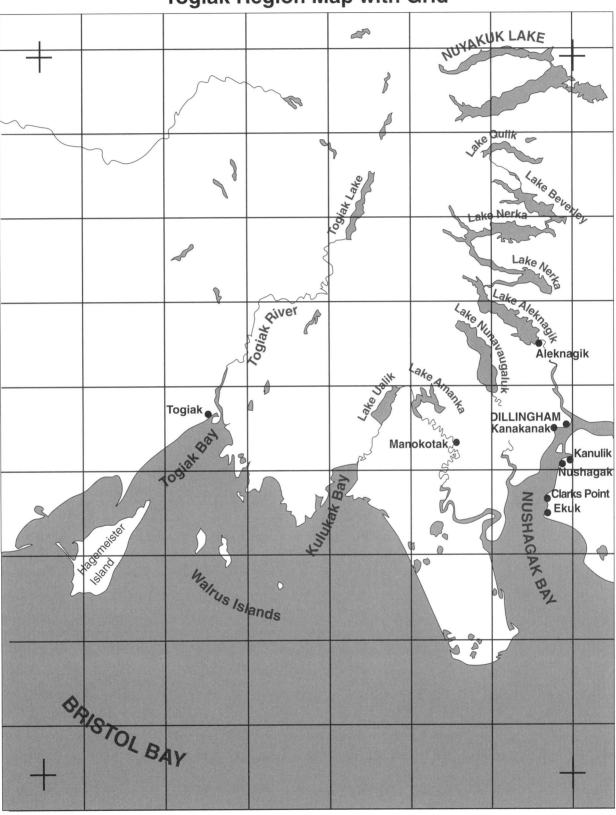

Blackline Master

Togiak Resource Mapping Game

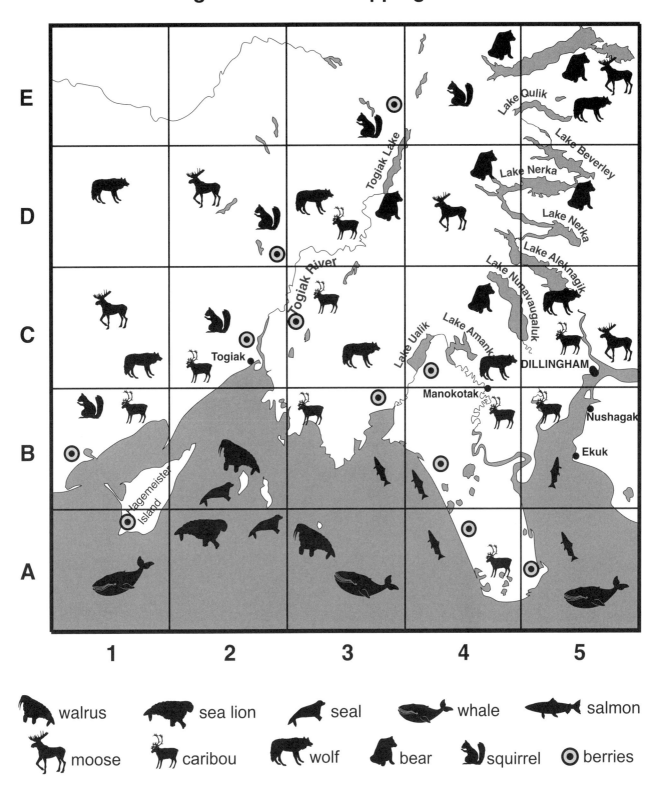

Blackline Master

Togiak Resource Mapping Game Cards

Blackline Master

Section 4:

Refining Conceptual and Practical Understanding of Measuring and Manipulating Shadow Data

Activity 11:
Exploring Measurement

In the previous activities, students have honed their observational techniques and their ability to record and analyze weather-related data, and they have applied these skills to make connections between math concepts and berry growth. In this activity, students pose a hypothesis that leads them into a series of inquiry-oriented investigations related to measurement concepts. Students use their shadows as an indirect way of observing changes in sunlight and its effect on plant growth.

As students measure their shadows, various misunderstandings about measuring arise. Does each group use their unit correctly, without skipping spaces? Do they start at the correct starting point? Do the units have gaps between them? Do they overlap? Encourage students to work through questions like these and allow group discussion to reach a consensus.

Part 1: Creating Units

The biggest challenge students face in this part of the activity is creating their own personal units. Concepts of unit size require complex thinking. Not only do students need to adhere to the measuring concepts they've been learning, but they need to use their judgment to create an appropriate unit for the task at hand. For example, if students use a five-foot-long unit, they would not acquire accurate information (if it was not further partitioned). Conversely, students might use a very small unit that would increase the probability of making counting or measuring mistakes by resulting in a large value.

Some teachers have questioned the introduction of personalized, nonstandard units. A teacher from North Pole remarked:

At first I wondered why you would want me to go backwards for instructing the measurement process. I already taught them standard measurement. Isn't it more important to teach them to use standard measures rather than making up their own? But then I discovered that they really didn't understand measuring at all. They had no understanding of units of measure. I discovered they were memorizing the vocabulary words, but not the meanings. They really had no grasp of the concepts of measuring. I watched them repeatedly begin measuring from the number one on their rulers, and the whole first inch was lost. They left gaps between the rulers and had no idea how to deal with the fractional units, even though the inches were clearly marked. (Rebecca Adams, personal communication, July 2003.)

Students will create personal tape measures using nonstandard units of their own choosing. In addition to solving the problem of unit size, students need to iterate their units. And, when the length they are measuring ends before a unit does, they will again face the issue of partial or incomplete units.

Goals
- To introduce nonstandard Yup'ik body measures
- To apply Yup'ik body measures to measuring classroom objects
- To create a personal tape measure by using an appropriate unit and by repeating that unit
- To solve the problem of partial or fractional units when encountered during measuring
- To apply the rules of measuring: equal-sized units, no overlapping or space between units, etc.

Materials
- *Berry Picking* storybook
- Adding machine paper
- Butcher paper
- Math Tool Kits
- Nonstandard measurement tools (paper clips, unifix cubes, string, books)
- Poster, Yup'ik Body Measures
- Student journals

Duration
Two class periods

Vocabulary
Ascending Order—a way to organize data from the lowest to the highest number

Descending Order—a way to organize data from the highest to the lowest number

Preparation

Have materials (adding machine paper) available for you and your students to make your own personal tape measures during the course of the lesson, or have one completed to show the students during steps 3 and 4. Your personal tape measure should also include partial units.

Activity 11: Exploring Measurement

Using adding machine paper, prepare several "personal measuring tapes" containing purposeful errors. One tape will demonstrate the error of unequal units, one with overlapping units, another with gaps between the units and one with gaps at the start and end points (see Figure 11.1).

Use a wall where the students' height measures can stay for approximately three to five days. Cover a section of the wall from the floor up with butcher paper so that students will be able to document their heights and their measurement strategies by writing on it.

Post the Yup'ik Body Measure Poster in a prominent place in the room.

For additional classroom ideas, see Susan Wainwright's chapter in *Investigating Real Data in the Classroom: Expanding Children's Understanding of Math and Science* (Eds. Lehrer and Schauble), pp. 55–62.

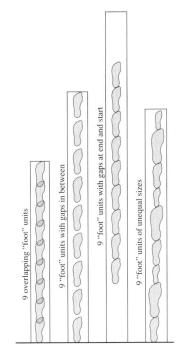

Fig. 11.1: Personal tape measures with errors

Instructions

1. Complete reading the *Berry Picking* story, page 73 to the end. Discuss the ending, characters and favorite parts.

2. Introduce Yup'ik Body measures to the class. Show the students the Yup'ik Body Measures poster. Practice making several of these measurements as a whole group. Then have students practice and explore these measures with a partner by measuring two or three items in the classroom.

3. Have students create a personal tape measure (see Figure 11.2). During this task, they may work in groups or in pairs. They need to choose their personal units; they may use any unit for their measuring tape, such as Yup'ik body measures, paper clips, unifix cubes, string, or a book. Make a variety of objects available.

4. Hand out adding machine paper and have the students choose an appropriate length for measuring their heights. Ask the students to use their unit to partition the adding machine paper (see Figure 11.2).

5. Students will then share their personal tape measures with the class. Ask, "Why did you choose this unit? Is it easy to use for this task?" Have students compare their tape to those of their classmates. These comparisons will help them fix mistakes on their tape measures and in developing conjectures later in the lesson. Teacher observation is critical at this point. Guide the students to note errors of spaces between units, unequal units, overlapping units, and gaps at the starting and end points.

Fig. 11.2: A personal tape measure

Teacher Note

When students measure their heights, measuring is done with a purpose; creating a unit for a personal tape measure requires the students to coordinate the units with the task. This process includes estimating.

6. When students have completed their personal tape measure, increase student understanding by posting one of the tapes you made earlier to demonstrate one of the common errors. Discuss each demonstration tape.

7. Students will measure their heights with their personal tape measures. How do the different groups of students solve the problem of partial units? Pose a question or solicit from students; what can we do to describe a person's height more accurately when it cannot be counted with whole units?

8. Have students share their solutions to the problem of partial units. Have them decide which solutions make sense. [**Suggestion:** Folding the last unit on their tape measures into halves or quarters, is one solution to the problem of fractions. Students would also need to name these fractional units and notate them.]

9. Have the students carefully label their adding machine paper with their names and heights. Let them post their height data on the wall. Observe the way students post their tapes to the wall. Do they leave a gap at the floor? Are they slanted?

10. Ask, "How can we use the data recorded on the adding machine paper to compare the students' heights?" "What could we add to each tape to know how many units tall you are?" This should lead students into the idea of writing numbers on their tapes for each unit drawn. This might provide a basis for discussing similarities between many different units of measure or common units of measure (see Figure 11.3).

Fig. 11.3: Student comparing classmates' heights

Part 2: Interpreting Height Data

In Part 2, students record their height data onto graph paper. This scaling down data from the wall onto paper challenges students to engage in proportional thinking. As students graph their measurements, they continue to examine the related concepts. The struggle to create an accurate data representation challenges them to consider scaling down when using non-common units.

Students will then be asked to transfer their graph paper representation into their journals. Assessing these graphs in the journals provides additional opportunities for insight into their mastery of the concepts of units and measurement.

The inverse relationship between the size of a unit and the number of units in a given measurement often confuses students. A small unit of measure will have a higher number value than a larger unit used to measure the same

Activity 11: Exploring Measurement

object. Students may think that the higher value is longer than the lower value because they do not associate the value with its unit. Activities like this enable students to work through these misunderstandings.

Students also confront the issue of nonstandard units and how to convert from one system of measure to another. This conversion process lays the groundwork for understanding the need and convenience of using common, standard units of measure.

Goals
- To pose and answer questions from the height data that students gathered
- To scale down physical height data and re-create in journals
- To realize that different units of measure will result in different numbers
- To understand that the number of units tells only how many times that unit is used

Materials
- Graph paper
- Crayons and pencils
- Math Tool Kit
- Student journals
- Height wall display from yesterday's activity

Duration
One class period

Preparation
In the next activity, students get a clearer understanding of the difficulties presented by not having common units of measure. Students will go back out to the height wall, question their data, and attempt to graph the heights of the members of their small group from yesterday. Students will need crayons and pencils and a hard surface to write on, and then they will need their journals to transfer their graphs into.

Instructions
1. Now that the height data on the wall looks like a graph, have students pose their conjectures or questions. They may ask questions such as, "How much taller is A than B? Are boys taller than girls?" Have the students pose questions of this data and allow others to try to answer.

2. Have students work in groups and answer the questions. Have the class discuss the responses.

3. Have students work in groups of three or four. Hand out graph paper. Have each group of students transfer their height/wall graphs onto the graph paper. Observe the way students represent their height onto the graph paper. Some may only pay attention to the number of units, others to the size of their unit and not to what the unit actually represents.

4. Discuss the difficulties encountered when scaling the graph down. Was it difficult to maintain proportions?

5. Have students glue their graphs into their journals.

Teacher Note

Students can tell visually who is the tallest or the shortest student, but when it comes to comparing across personal units, it is very difficult to make sense of these measures. This is a good place for students to understand and see the need for more common units. For example, Jerry is two units tall and Barbara is four units tall. Who is taller? (See Figure 11.4.)

Teacher Note

Typically students may just transfer the number of units to the graph paper without regard to the unit's relative length. As students transfer their physical data onto graphs, they have to reduce or scale it down. In the next lesson, students will work through the idea of common units which will make data (heights) easier to compare.

Jerry **Barbara**

Fig. 11.4: Interpreting height data. Who is taller, Jerry or Barbara? Students may say "look, Jerry is bigger," or some may say "Barbara is 4 and Jerry is 2." Have students reconcile their conflicting understandings.

Activity 12: Measuring Shadows

In previous lessons, students have measured their heights using their own personal tape measures. In doing this, students gained more experience partitioning a length into equal units and working with partial units.

Using personalized units presents certain limitations in comparing the values of lengths measured with different units. Students will realize that converting these units to some common unit is cumbersome. The *Big John and Little Henry* story also points out some of the limitations of nonstandard or personalized measures.

Part 1: Finding Common Units

This activity marks the transition from using nonstandard units to common units that can be more easily communicated to others. Students may suggest the use of a ruler or tape measure, and this activity is a good place to include these suggestions. However, students may choose to have the class use a common unit such as a Yup'ik body measure, a book, an eraser, etc. This would be an intermediary step between personalized units and standard units such as a store-bought ruler or tape measure.

Students once again face the issue of partitioning and partial units as they rework their personal tape measures into common units. Measuring is almost always an approximation; students should think through the problem of sizing a partial unit to provide sufficient data for a meaningful measurement value. Again, students face key issues of measurement, such as starting point and consistent units.

They continue to work on their height graphs, making the graphs more understandable to others outside of the class. Students will have to represent their height data in a way that makes their idiosyncratic, personalized measurements more generally understandable.

> **Cultural Note**
>
> When berry picking on hot days, Yup'ik pickers sometimes wait for clouds to come over and shade the area. Not only is it cooler then, but it's easier to see the berries without the glare of the sun.
>
> The following story was related by Mary George of Akiachak, Alaska:
>
> *One day, a lady was picking berries near a mountain and the sun was bright. She noticed two dark things moving along the ground. She got scared, thinking they might be bears. But then she looked up and noticed two very pretty clouds moving along under the mountain, which were casting the shadows on the ground.*

Goals
- To measure height tapes, using a common unit
- To record their measurements onto their height tapes, using a common unit
- To pose questions and use their data to answer their conjectures accurately
- To use the collected height data to create a graph (labeled, units, and equal spacing)

Materials

- *Big John and Little Henry* storybook
- Graph paper
- Math Tool Kits
- Personal tape measures from the height wall
- Poster, Yup'ik Body Measures
- Students' journal entries from the previous lesson
- Balls of string—all one color to represent student heights
- Paper clips

Duration

One or two class periods

Vocabulary

Conjecture—a prediction based on incomplete data

Malruk Naparneq—a Yup'ik length of measurement that is the same as two fists with thumbs extended and touching

Nonstandard Measure—a measure found by using a unit that differs from one person to another, such as using a body measure instead of a ruler

Ascending Order—a way to organize data from the lowest to the highest number

Descending Order—a way to organize data from the highest to the lowest number

Preparation

Post the Yup'ik Body Measures Poster. Have the work from the previous lesson available. Also, have nonstretching string available that students will post on the height wall chart as they remove their personal tape measures. These strings should all be the same color, but it will need to be a different color from the considerable amount of string the students will use to represent their shadow lengths for Part 2. You will need more of the shadow string color than the height string. Using two different colors should help to keep shadow and height information separate. You will need a copy of the story, *Big John and Little Henry*, to read to the students.

Instructions

1. Read the *Big John and Little Henry* story. Have a class discussion about the problems the characters encountered when measuring.

Activity 12: Measuring Shadows

2. Have a student volunteer to have his height measured with the Yup'ik body measure *malruk naparneq* (two hands in a fist, thumbs outstretched and touching; see Figure 12.1). Have another student use this unit to measure the volunteer. Write the number of units on the board. Ask for other students to repeat the procedure, writing the number on the board each time. Students may arrive at different numbers of units even though they are measuring the same height.

Fig. 12.1: Malruk naparneq

3. Discuss the reasons for the differences.

4. When the students understand the need for a common unit of measure, have them discuss and decide which will be the common unit for the class. After the discussion, have the students remove their tape measures from the height wall. Have them turn the paper over and repartition it according to the new common unit.

5. Have students return to the height wall and post this side of their measuring tapes. Have them number their units once again. Then share strategies and solutions for partial units while others look on.

6. Students should write their heights in this unit on the height wall above their names. Continue to facilitate the process of making this "graph" more understandable for anyone who walks by.

7. Discuss the revisions that the students are making to their height graphs. Facilitate discussion and students' solutions to issues of common units, partial units, labeling, and titles. Ask for student volunteers to write these parts of a graph on the wall chart.

8. Have the students move back into the same small groups as in Activity 11, Part 2. They should take out their journal entries from the previous lesson. Hand out one sheet of graph paper per group and ask students to graph their group's revised heights using this new common measure.

9. Facilitate the students' discussion on the differences between using personal measures and common units.

10. Students may now go to the height chart with balls of string (all the same color). They will need to measure and tape up a string to represent their height as they remove their personal tape measures from the wall. They will need these tape measures for tomorrow's shadow activity. Have students carefully roll up their personal tape measures and secure them with a paper clip before putting them in their tool kits.

Teacher Note

When using the same body measure, different people will have different lengths for the same unit, just like in the *Big John and Little Henry* story. This difference exposes both the strengths and limitations of body measures: they are easily usable but not easily comparable. Students will probably come up with confusing comparisons that should allow them to see the difficulty in comparing and communicating dissimilar units. Having the students work through this comparative process is the key to understanding the need to measure with common units.

> **Teacher Note**
>
> To make the height data meaningful, students need to establish a common unit so that they can compare heights at a glance. Once the data is organized, it becomes a useful collection of which students then can ask questions. For example, "Who is the tallest student? Are boys taller than girls? How many people are taller than...?" This lesson parallels the earlier lesson on categorical graphs without numbers, since the students will need to work through key graphing issues such as starting at the base line. Some students may start at zero while others may start at 1; some may start at the top and work down while others work up.

Part 2: Measuring Shadows and Answering Simple Conjectures

The ability to note changes in a person's shadow—its length and direction—is a fundamental observational skill often used to tell time and direction by Yup'ik elders as they travel across the tundra.

Sunlight not only provides guidance to know time and direction, but sunlight is also a major factor in plant growth and health. In this part of the activity, students will measure their shadows to learn about light at different times of day. Students will be curious about when their shadows are the shortest or longest and what effect the position of the sun has on their shadows.

Considering the size of the state of Alaska and the great variation in daylight hours and the position of the sun in different areas of the state, not to mention daylight savings time, the length of a person's shadow at noon will vary by location, time of year, etc. Students in a far western part of the state were surprised to see their shadows continuing to get shorter at 1 p.m., 2 p.m., and even later. They asked, "When will it be the shortest?" Making conjectures, gathering evidence, representing the evidence, and interpreting data are all important components of today's lesson.

As students engage in the process of collecting and recording the data, they will organize their data into tables. They also have additional opportunities to measure accurately. Using string or their common tape measure, they continue to partition a unit (the string) into common units and iterate (repeat) the common unit to the full length of their shadows. Students answer their own questions and ask new questions of the data they collect. They can physically manipulate their data by arranging it in ascending or descending order according to length, and they line up in height order next to the display of shadow data. Does the class height order match the shadow data? If so, ask why; if not, have the students explore the discrepancy.

Goals
- To measure and record their shadow measurements using a common unit
- To determine accurate starting and end points and partial units
- To use collected height and shadow data to develop conjectures
- To use collected height and shadow data to prove conjectures

Materials
- *Bear Shadow,* by Frank Asch (or other shadow-related storybook)
- Balls of string—all the same color, but a different color from the previously used height string

Activity 12: Measuring Shadows

- Student journals
- Masking tape
- Butcher paper
- Math Tool Kits

Duration

Two class periods

Vocabulary

Conjecture—prediction based on incomplete evidence

Common unit of measure—any unit of measure, standard or nonstandard, chosen by a group as the agreed-upon unit of measure for a particular purpose or at a particular time

Shadow—the image cast by an object that is blocking a light source

Preparation

Remind the students that they have been collecting weather data and have learned how weather affects plant growth. They have grown plants in class, paying attention to how much water the plants need. Now they will be observing their shadows as a way to observe sunlight, the last factor related to plant growth that this module covers. Students will use their personal tape measures and the chosen common unit to measure their shadows. They will record their measurements and observations in their journals. Students will need the new string colors, masking tape, their Math Tool Kits, and their journals for this lesson.

Instructions

1. Let the students know that during the next few days they will be going outside to investigate their shadows. Some elders use their shadows to tell time and to find their way home.

2. Read *Bear Shadow* (if available) to the students. Students may want to discuss what causes shadows, what makes them long or short, and how shadows change.

3. After the discussion, ask, "Is your shadow longer, shorter, or the same as your height?" Write down the students' answers. Ask the students how they can answer this question. (If possible, follow the students' suggestions for answering this conjecture. Otherwise, follow the steps below.)

Making Conjectures

Teacher Note

Becky Adams' students in North Pole, Alaska, went outside and immediately began comparing their shadows. "Whose is the tallest?" Some students ran ahead of the others and declared their shadows were the longest. Because the students did not have a common starting point, there was confusion. The students who were saying, "Mine is longer than yours," thought this was true because of where the shadow ended, even though its actual length was shorter. In fact, this statement came from one of the shortest students in the class. The next day everyone was highly focused on the starting point. Adams said, "We made a line with colored chalk, put our heels at the line and everybody measured his or her shadows from there."

4. Tell the class that they will go outside to measure their shadows at 10 a.m., noon, and at 2 p.m. to find out if their shadows are shorter, taller, or the same as their height at these three times. The students will use their personal tape measures for this task.

5. Demonstrate the first conjecture, that their shadows will be shorter than their actual heights. Explain what is meant by a conjecture. The initial conjectures will assert ideas that you want to explore about the shadow data. Each student (or group) should now write a conjecture in his or her journal before measuring the shadow. Have students share their conjectures. These need to be reasonable conjectures answerable through measuring data collected by measuring shadow lengths. Choose several of the conjectures to write on butcher paper to discuss following this activity.

6. Students will plan strategies for measuring their shadows. Ask, "Where does the shadow begin? Where do we start to measure?" Discuss their approaches, and if possible, have the class agree upon a method of measuring their shadow lengths.

7. Divide students into groups. Students will need their math tool kits and journals and will share balls of string (of different color than their height string). They will also need a piece of masking tape to wrap around their shadow string for labeling.

8. At 10 a.m., go outside. In groups, students can mark starting and ending points of each other's shadow on the ground. Cut a string to match this length. Using the tape to wrap around the string, students will label it with their name and the time.

9. After shadow strings have been cut and labeled, students will need to use their personal measuring tape to determine how many of their common units long this string—their shadow—is. This information should be recorded in their journals.

10. Repeat step 8 at noon and again at 2 p.m.

11. Ask students to look for similarities and differences in shadow lengths recorded for 10 a.m., noon, and 2 p.m. Have them examine the relationship between shadow length and the students' height. For example, they may consider length and direction. Can students make sense of the various comparisons? Encourage students to find more than one way of comparing the shadows.

12. Have each group of students answer their conjecture. For example, are their shadows shorter, taller, or the same as their heights? Also, discuss the conjectures written on the butcher paper.

Activity 12: Measuring Shadows

13. Encourage students to manipulate their shadow strings in various ways to answer the conjectures, such as in ascending and descending order. Allow them to decide how they want to post these shadow strings with their height string on the wall.

14. Have the students post their shadow data, well labeled (name, date, time), next to their height data on the wall. Have the class report their findings (read their data from the graph and compare their height and shadow measurements).

15. Ask questions of the data. This is an excellent time for the students to ask more questions of the data and to organize the data in order to answer their conjectures. For example, the students may ask, "Who has the longest shadow? Who has the shortest shadow? What is the relationship of height to shadow height?"

> **Teacher Note**
>
> Encourage students to conjecture about the data. For example, Sandi Pendergrast's class from Anchorage wondered when their shadows would be equal to their heights. This led to a discussion of what time of day they could measure shadows to determine the height of other objects around them.

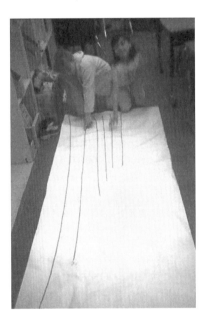

Fig. 12.2: Students compare their data

> **Teacher Note**
>
> Barbara Arena's students in Kotlik, Alaska, wanted to know when their shadows would be shortest. They decided to take multiple measurements between noon and 2:30 p.m. until they found when the two lengths were equal. Because they live in western Alaska and because Alaska observes daylight savings time, their shortest shadow fell at approximately 2 p.m.

Measuring and Number Line: Connecting Arithmetic and Measuring

One big concept for students to learn while measuring is to recognize the connection between measuring, counting, and number relations. This note extends the activities in this module by connecting measuring to arithmetic: two-digit addition and subtraction problems. A ruler is another representation of a number line, which students may not realize. Therefore, this note makes this connection explicit as a way to encourage students to use inventive algorithms as shown in Figure 12.3 below. For example, when students measure their height using a yardstick, they will use two yardsticks or the same yardstick twice. Two students can compare their heights by adding or subtracting. You can generate different math problems. For example:

1. Fred measured Bob and found that he was 46 inches tall. He measured George and found he was 48 inches tall. How much taller is George than Bob?
2. Fred knows he is 4 inches taller than George. How tall is Fred?
3. Sue is 8 inches shorter than George. How tall is Sue?
4. How many more inches does Sue need to grow so that she can be the same height as George?
5. My yardstick broke at 7 inches and I measured my shadow using the broken yardstick and it read 35 inches. How long was my shadow?
6. I used two yardsticks to measure my shadow and it read one yardstick plus 23 inches. How long was my shadow?

These problems could be thought of as adding or subtracting on a number line. Problems like this can be extended in the ways noted above and by having students generate their own word problems. Lastly, if students show their thinking on a number line or measuring device it can provide insights into their ways of grouping numbers.

For example, Bob measured his shadow at 3 p.m. and it was 62 inches long. He had also measured his shadow at noon and it was 37 inches long. How much longer is his 3 p.m. shadow than his noon shadow? Students can use the number line below as a way to help shape their math thinking. Also, students' solutions provide insights into their mathematical development—for example, do they use a tens and ones approach or do they count by ones? Multiple strategies are possible. Have students share their strategies.

Fig. 12.3: Number lines and subtraction

For additional details see the research of Koeno Gravemeijer, Janet Bowers, and Michelle Stephan (2003). A Hypothetical Learning Trajectory on Measurement and Flexible Arithmetic. *Journal for Research in Mathematics Education Monograph Number 12: Supporting Students' Development of Measuring Conceptions: Analyzing Students' Learning in Social Context*. Edited by Michelle Stephan, Janet Bowers, and Paul Cobb, with Koeno Gravemeijer.

Activity 12: Measuring Shadows

Part 3: Measuring Shadows with "Broken" Tape Measures

Goals
- To assess students' ability to measure with a "broken" tape measure
- To reinforce a zero starting point

Materials
- Balls of string—shadow color
- Math Tool Kits
- Student journals
- Masking tape
- Paper clips
- Butcher paper
- Student personal tape measures

Measuring and Starting Points

Duration
One class period

Vocabulary
Standard Measures—accepted ways of measuring using an established standard, for example, inches or centimeters for length and pounds for weight

Preparation
In this part of the activity, students will measure their last shadow string using a "broken" personal tape measure. You will need your own personal tape measure and a few paper clips available in case a student has lost theirs. First you will accept any new conjectures before taking the class outside for their last measurement. After returning to class, you will pose the questions leading to the use of a "broken" tape measure.

Instructions

1. Students will collect another shadow measurement at 2 p.m. Ask for any new conjectures and write them on the butcher paper as before. Take students out to cut and label their last shadow string.

2. After returning to the classroom, ask, "What would happen if my personal tape measure had the first two units accidentally torn off? Would I still be able to use it? Would I still be able to measure with it? Would it be accurate?" Discuss each question, and then give each student a

Teacher Note

Watch the students as they attempt to measure with their "broken" measuring tape. Where do they begin to measure? How are they counting? Are they starting with "one" or "two" or even "three"? Becky Adams of North Pole had one student who measured to the seventh unit, then added the missing two units and called his measurement 9 units long, even though it was only 5 units.

paperclip. Have students take their personal tape measures from their tool kits and fold over the first two units. Secure with paper clips.

3. Have students work in small groups and allow each student to attempt to measure their newest shadow string with their "broken" personal tape measures. Each student will need to write their answer in their journals with the rest of the shadow data.

4. Have each group share with the class how they solved the problems of measuring with a broken tape measure. How did they deal with partial final units? What did they do when their measuring tape was not long enough?

5. Have each student remove their paper clips and remeasure their latest shadow string. Discuss if their answers differ. Have students resolve the differences. Then post this string on the height wall with their other strings.

Part 4: Broken Rulers

Note that the following activity, "Broken Ruler," was derived from the work of Jeffrey Barett and Sandra Dickson and in consultation with Dr. Richard Lehrer.

Students have had much experience throughout this module developing their own units of measure (personal tape measures) and coming to understand the need for common, standard units of measure. However, as noted earlier in the module, students may mask their conceptual misunderstandings by being able to measure a distance accurately. To ensure that students are competent at measuring, the following activity provides them with a new challenge: to measure with a broken ruler. Classroom research (Barett and Dickson, 2003; Lehrer, 2003) suggests that when using a ruler, students may confuse the number on the ruler with measurement lengths. In other words, where does the count begin? Do students count spaces or numerals? Do they know what each represents? The intent of this activity is to illuminate hidden misunderstandings.

Students will use broken rulers of various lengths. Students may think that they cannot measure their shadows because the ruler is broken. But as students practice measuring with the broken rulers, they learn that they can measure without the number zero on their rulers, and that regardless of what numeral may be written there, the beginning of the first complete unit serves as a zero point. They may still be confused about where the count ends, but they will learn that it is actually the space between the numbers that is counted while measuring. Measuring with a broken ruler provides an opportunity for you to assess students' conceptual understanding of measurement.

Activity 12: Measuring Shadows

Goals
- To understand the need for standard measurement
- To measure height and shadow lengths using a standard measurement
- To record measurements using a standard measure
- To understand how to partition a standard measure

Materials
- Handout, Broken Ruler, one per group
- Student journals
- Student shadow and height strings on the height wall

Duration
One class period

Vocabulary
Standard Measures—accepted ways of measuring, using an established standard, for example, inches and centimeters for lengths, and pounds for weight

Preparation
Make enough copies of the Broken Ruler handout so that each student has one cut-out ruler. Students will need to be able to go to the height wall to measure their strings with broken rulers.

Instructions

1. Give one Broken Ruler handout to each small group. Have the students cut out each broken ruler and choose one. Explain that these are samples of broken rulers, then ask, "Can we still measure things accurately, even if all we have is a broken ruler?" Discuss this idea and allow for student reasoning to be shared with the class.

2. Discuss the importance of using a standard measure, like inches. How would this help people who are not from the classroom to understand our height wall better? Tell the students that they will need to use the broken ruler that they choose to measure their height and shadow strings with inches. They will need to write this new measure above each of their strings and in their journals. Have students within each group check their height and shadow measurements recorded from using a broken tape measure. Have them resolve discrepancies.

Teacher Note

Observe students using their broken rulers. Do they realize that they can still measure with a broken ruler? Do they begin measuring with the first number or after the first space? Assist students as they measure.

This is the first time in this module that students measure in inches. However, because they are using a broken ruler, this is a good assessment if students are able to accurately measure.

3. Watch the students closely during this process to assess for understanding. If your students have too much difficulty with this task, you may choose to end this activity early and proceed to the extended version of the "Broken Ruler Activity" (see page 123).

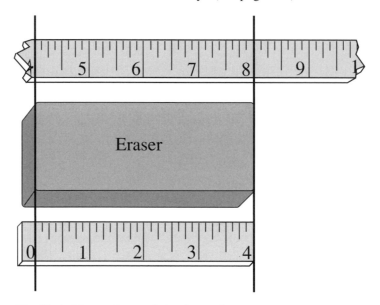

Fig. 12.4: Measuring with broken rulers

Part 5: Continuing to Answer Our Conjectures; Finding Patterns in Data

Students begin to make connections between the position of the sun and the length of their shadows. Students can observe patterns in the data, particularly if they collect the data over time.

They can observe that shadow length and direction change daily. Students observe that as fall becomes winter, shadow lengths recorded at the same time daily will increase. They will realize that the temperature falls as the shadows lengthen and the process reverses as spring moves toward summer. This activity concludes with students interpreting their data and relating it to their experience of the environment and seasons.

You have the opportunity to assess students' understanding of the relationship between shadow lengths on different days and their ability to work with more abstract data.

Goals
- To organize data in a meaningful way
- To prove conjectures using collected data
- To manipulate and interpret string data to answer conjectures

Activity 12: Measuring Shadows

Extended Practice Measuring: "Broken Ruler Activity"

This activity is designed to give your students experience using standard measures. It functions both as an assessment and an opportunity for your students to integrate their knowledge of measuring. They have experienced measuring with nonstandard units and common units and have discussed many of the problems inherent with developing measurement skills and concepts. This final measuring activity places its emphasis on the use of the zero point, and that it is the space between the number that matters.

Goals
- To assess student progress with correct ruler use
- To reinforce the measuring process using spaced units rather than numbers
- To improve math communication skills
- To reinforce a zero starting point

Materials
- Student journals
- Math Tool Kits
- Worksheet, Broken Ruler Worksheet, one per student
- Handout, Broken Rulers, one per group of students

Duration
One class period

Preparation
Make enough copies of the Broken Ruler handout so that each student has one cut-out ruler. Also, students will each need a copy of the Broken Ruler worksheet.

Instructions
1. Pass out the Broken Ruler handouts, one to each small group. Have students cut out each broken ruler and choose one.
2. Give each student the Broken Ruler worksheet and ask them to work independently to answer the problems. As you walk around and monitor student work, observe how students use their broken rulers to measure. Do the students count the spaces between the numbers as the actual inches? Can the students discover where their "zero point" is?
3. You may choose to collect the worksheets for grading purposes. Bring the whole class back together to allow students to discuss and correct any errors. Have a student volunteer show the class how he or she used the broken ruler to find the length of the first line. Have another volunteer who is using a different broken ruler also demonstrate for the first line on the worksheet. Repeat this process, allowing students to remeasure and correct their answers as needed to complete the worksheet.
4. Students will glue this worksheet into their journals as they discuss what they have learned.

Teacher Note

Barbara Arena, a teacher from Kotlik, reported that her students placed their shadow data in the ascending order. She had her students stand in order by height to determine if they matched their data. The students noticed that some of the shorter children actually had longer shadows. Instead of deciding to measure again, the students simply changed their order! This leads to a discussion of accuracy in measurement and working with data. This is a good opportunity for students to see if their conjectures are supported by the data.

Fig. 12.5: Students compare their data

Materials

- Long sheets of butcher paper, one per student
- Student journals
- Scotch tape
- Math Tool Kits

Duration

One class period

Instructions

1. Ask students to take out their journals and locate their shadow questions. Ask for volunteers to read their conjectures or questions. Ask, "Can our data answer these questions?" Students should have questions such as, "Was my shadow longer in the morning or at noon? Who has the longest shadow?"

2. Facilitate learning by asking, "How can we organize our data (height and shadow strings) to answer these questions?" Allow students to set up their graphs (see Figure 12.5).

3. Have each group of students move the data around, for example in ascending or descending order if that responds to their questions. See examples in Figures 12.6 and 12.7. You may need to model one example of manipulating the data so that students understand that they can manipulate this physical data and make a variety of comparisons. This is an important synthesis activity for the students since it allows them to work with sophisticated data in a concrete way.

4. After the data is reorganized, ask what they now notice and what questions they can answer.

5. Have students make a journal entry. Students will choose one conjecture from either those they have written previously or from those that the class has just discussed. The students will answer this conjecture by providing data in the form of a drawing and a written description. For example, some students wondered when their shadow would be the longest.

6. As the culminating shadow activity, students will build their own graphs using their own data. They will need their height data and the data from previous shadow measurements. Each student will transfer this onto his or her own sheet of butcher paper. Students will label times, height, and scale (units) and draw in the relative position of the sun to

Activity 12: Measuring Shadows

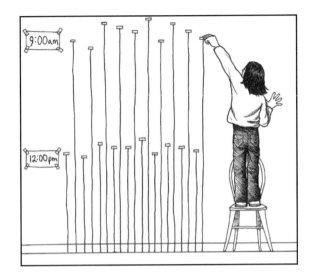

Fig. 12.6: Height data organized by students' shadow at noon and 9 a.m.

Fig. 12.7: Height data organized by time

the corresponding shadow (data). Note: the lower the sun is in the sky, the longer the shadow that is cast.

7. When completed, the graph should show the height of the shadow and the sun's position. This may be displayed throughout the school or used for decorations for the Family Berry Feast the class will soon be having. By using their graphs, the students can show the relationship between the time of day and their shadow lengths. Seasonal changes in the position of the sun on the horizon impact the plants' growth, while the students' graphs will show daily variations. Eventually, the students can roll up the graphs, band them, and take them home.

Broken Ruler Worksheet

Name _____   Date _____

1. How long is your ruler?

2. Use your broken ruler and measure the following lines. Write the answer in the space provided.

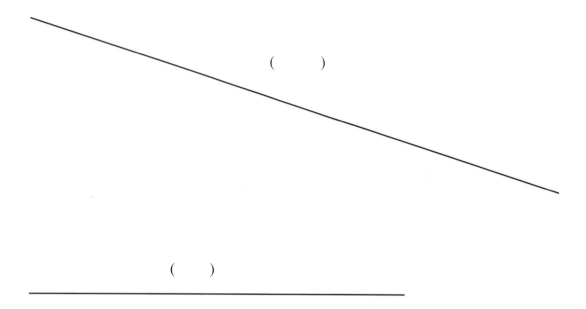

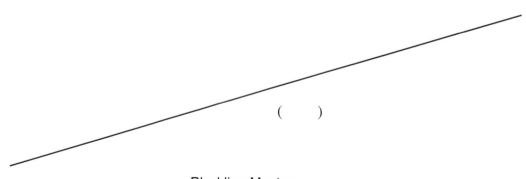

Blackline Master

Broken Ruler Handout

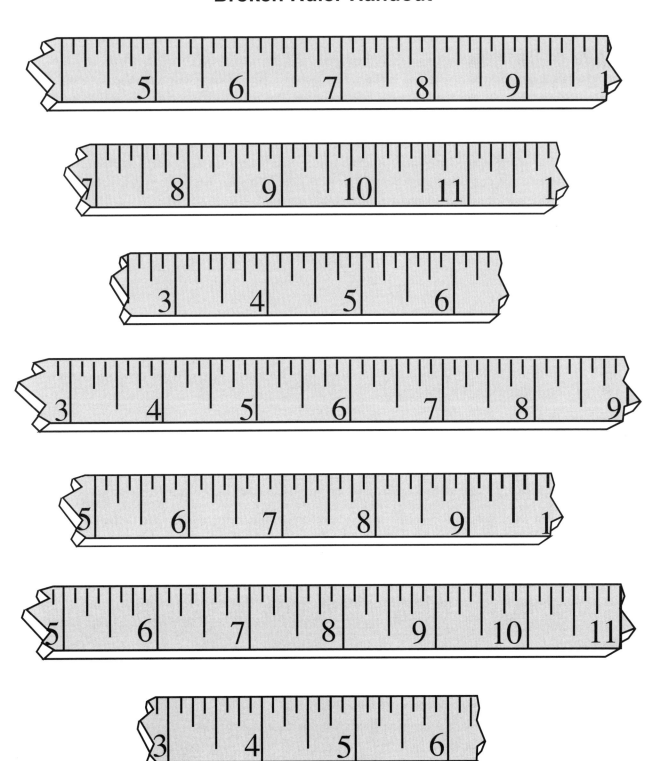

Blackline Master

Cultural Note

Uses of Berries

Not to waste anything or take more than is needed is a practice and a value among most Yup'ik people. Many stories told by the Yup'ik elders confirm this practice. Wassily Simeon tells us, "We never picked too much. Always just right. Just as much as we think we need." An occasional unripe or rotten berry might stray into the bucket, and it will be discarded.

If people get an excess supply, they share it with others, or use the excess for barter. Both Samuel Ivan and Evelyn Yanez said that they would always give their extra berries to those who were unable to get their own, regardless of the reason. Nastasia Wahlberg discusses trading:

> *Sometimes we would do exchanges. Sometimes I didn't get enough crowberries, but I would trade some moose meat for crowberries. Sometimes we would trade for seal oil.*

Dyes

Nowadays, it is uncommon to use berries for dyes, because it is much easier to buy and use commercial ones. However, in earlier times berries were used to color different things. Anuska Petla of Dillingham remembers heating up cranberries to dye walrus intestine. The dried intestines turned dark pink from the cranberries, while crowberries turned them purple, and blueberries made them blue. She adds that the dye kept the same color as the berry on the bush.

Medicines

Cranberries (see Figure 12.7) were used as a remedy for many ailments, including colds, diarrhea, and canker sores. "Cranberries are angels," Anuska Petla says with sincerity. She explains that if two spoonfuls of crushed cranberries are taken at night, one would wake up the next day and the mouth sore would be gone.

Lillie Gamechuk from Manokotak remembers that "a drop of cranberry juice was used for pink eye." Sassa Peterson from Manokotak, mentions that the stems of cranberries can be boiled, and the steam breathed in would relieve those suffering from asthma.

However, some berries are believed to carry health risks. For instance, it is believed that cloudberries should not be eaten after surgery, as they inhibit the healing process and increase the chance of infection.

Foods

Yup'ik Eskimos still rely on berries as a main food source. Berries are often eaten plain, but they can also be prepared as jams, puddings, or numerous other dishes. Margaret Wassillie likes to save overripe salmon berries, freeze them, thaw them, and use the juice for pudding.

By far the most popular way to eat berries is as *atsiuraq*, a mixture of berries and sugar, fish eggs, deboned fish, mashed potatoes, and seal oil, among other ingredients, cooked to the

Fig. 12.7: Low bush cranberries

consistency of pudding. The elders alter their recipes to suit the availability of ingredients in their area. Some of these recipes can be found below.

Nastasia Wahlberg mentioned the earliest *akutaq*, "Eskimo ice cream" (also called *uqiinaq*), and how it can serve as a survival food by providing an instant warming effect:

The first akutaq we made from cloudberries. We crush them really fine. We mix seal oil, sugar, and then it is ready to eat. It is potent, and it can make you sweat. It has a lot of acid. You feel a tingle from it.

Traditional Yup'ik Eskimo Recipes

Most Yup'ik measurements are expressed informally, such as by handfuls rather than as cups or liters. Much of the mixing is done with one's hands. Samuel Ivan emphasizes, "There is no set measurement, just estimation."

Where we use standard measures in the traditional recipes that follow, please note that these are rough approximations for the Yup'ik measures.

A *passin,* or an instrument used to crush berries (see Figure 12.8) is used for the following recipes.

Aapengayak: A Traditional Berry Dish by Anuska Nanalook

Place king salmon, red salmon, or silver salmon eggs in a small barrel underground, or in a shaded pit, to ferment for about one week. When ready, crush approximately ½ cup of fermented salmon eggs. While stirring, add ½ cup of seal oil (or Crisco) in intervals. Keep mixing (approximately 300 mixes). Add ½ to 1 cup of sugar. When the fish eggs, seal oil, and sugar have thickened, add about ½ gallon of crowberries (sometimes known as blackberries).

Fig. 12.8: A passin, *or berry crusher*

Qerpertaq (Variation from the above recipe, author unknown):

Using your clean bare hands, smash pike eggs, mix in seal oil, add sugar, and then add cranberries.

Contemporary *Akutaq* by Nastasia Wahlberg and Grace Gamechuk

- 2-gallon plastic bag of berries (any kind of berry)
- 3 to 4 heaping tablespoons of shortening
- 1 tbsp. vegetable oil
- ⅛ to ¼ cup of water or berry juice
- 1 cup of sugar

Mix the above ingredients (except oil and sugar) with hands until fluffy. Add 1 tablespoon of vegetable oil. Stir to make lighter and more slippery. Add 1 cup of sugar and thoroughly mix.

Enjoy your first Eskimo ice cream!

Blueberry *Akutaq* by Grace Gamechuk from Manokotak [*Manuquutaq*]
(Follow instructions from *akutaq* recipe above but replace water with applesauce)

- 1 gallon of blueberries
- 1½ cups of shortening
- 1 Tbsp. vegetable oil
- 1 to 1½ cups of sugar
- ½ to 1 cup of applesauce

Salmonberry *Akutaq* with Greens and Applesauce by Nancy Sharp from Manokotak

- 1 gallon of salmonberries
- 1 cup of sugar
- 1 cup of shortening
- ½ to 1 cup of applesauce
- ½ cup of green sour dock

Stir or whip the shortening until it fluffs. Add sugar and whip the mixture again until fluffiness increases. Whip in the applesauce and fluff it up. Add the green sour dock. Add the salmonberries. Eat and enjoy.

Fig. 12.9: A cup of berries

Cloudberry and Crowberry *Akutaq*

- 1 gallon of cloudberries
- 1 quart of crowberries
- 1½ cups of shortening (can substitute instant mashed potatoes)
- 1 cup of sugar
- ½ cup of evaporated milk or water

Mix ingredients until fluffy. Chill.

Blueberry, Crowberry, and Potato *Akutaq*

- 1 ½ cups of blueberries
- 1 gallon of crowberries
- 1 handful of shortening
- 2 cups of dried mashed potatoes
- 1 cup of sugar

First, mix dried mashed potatoes with hot water until fluffy and set aside until cool. In a large bowl, mix Crisco shortening, sugar, and mashed potatoes until the mixture is light and fluffy. Mix in thawed out crowberries. Add blueberries and mix into *akutaq*.

Fig. 12.10: Making akutaq—*Eskimo ice cream*

To avoid a greasy texture with your *akutaq,* try this helpful hint from Linda Brown of Ekwok. Linda says if you are using berries at room temperature, you will also need a small amount of water to blend in with the shortening to make it good and fluffy before adding the berries. If you are using cold or frozen berries, you will need a small quantity of vegetable oil to fluff up the shortening before adding the berries.

Activity 13: Using the Berries and Preparing for the Berry Family Feast Day: Students Predict the Next Berry Season

In this activity, students will invite elders, parents, and siblings into their classroom for a feast and a presentation of their math projects. You may choose to design this activity to your area and your group of parents. You will want your students to draft a letter of invitation. Ask your students for a list of food items that contain berries and ask for volunteers to bring these items in. Students may want to decorate the room and the hall with all of the math work they have collected from the *Picking Berries* module lessons. Students will have their collected weather data. They will predict the next season's berry harvest based on their data.

Before parents arrive, students will work on one last math lesson. Students will need to figure out how to double or triple an *akutaq* recipe so that there will be enough for all to try at the feast. Students will use this opportunity to further explore the use of standard and nonstandard measurement.

You may choose to begin the first part of this activity a week or so before the Berry Family Feast, but save the actual making of the *akutaq* for the day that their visitors will arrive to ensure it is tasty and fresh. You also may use art lessons or other class time to integrate the writing of the invitation letter. Read through these last two activities and decide what would best fit your needs and teaching style.

Part 1: Feast Day Preparation

Goals
- To practice presentation skills and use of math vocabulary words

Materials
- Math Tool Kits
- Student journals
- A *passin,* or berry crusher (see Figure 13.1) (optional)
- All of the graphs, charts, conjecture displays, and other materials from the module

Duration

One class period

Vocabulary

Passin—a traditional native berry crusher

Instructions

1. Share with students the information you gathered from the cultural notes about the various uses for berries, including berries used as a coloring dye for grasses and clothing. Demonstrate if you wish. If you own a berry crusher (*passin*), demonstrate its use. If not, share Figures 13.1 and 13.2 and explain.

2. Tell the students they are going to have a Berry Family Feast (or the class may choose their own title). Determine how you want to organize your party and tell the students, "To make this a really good feast, we are going to invite elders, parents, and little brothers and sisters who aren't in school yet. We'll want our visitors to eat all kinds of different things made with berries, and we'll want to entertain them by sharing all of the math projects we've been working on during the *Picking Berries* math module."

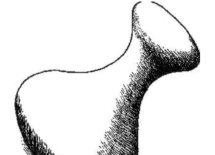

Fig. 13.1: Passin, *or berry crusher*

Fig. 13.2: Girl making akutaq, *Eskimo ice cream*

Activity 13: Using the Berries and Preparing for the Berry Family Feast Day

3. Tell the students that they will need to formally invite their guests by sending them invitation letters.

4. Discuss the invitations and what information will need to be included so that the visitors will know what to do. Write this information on the board and ask for volunteers to make sample invitations.

5. While the volunteers are working on the invitation samples, ask the rest of the class to think of all the food items that are made with berries. Write these on the board. Ask if these are the kinds of food they want to have at their party. (You will need to decide how to deal with the food situation. You may bring all of the food, ask for donations of these items, assign various items to each family to be responsible for bringing in, or even ask for students to volunteer their family to bring a specific item, etc.)

6. When students finish their invitation samples, share them with the class for a discussion. You will need to decide whether the class will be voting on one or two of the designs to make copies of for everyone, or if you will take all of their ideas to make a master copy for each student to color and take home.

7. Students will need to color their invitations for home and include the item that their family is expected to donate, if any. These will need to be taken home several days in advance, so that families may plan their visit and their food item.

8. Bring out all of the graphs and charts that the students have worked on. You may even want to include the maps and transparencies that students have been studying.

9. Based on the collected weather data, have each group "predict" the next season's berry harvest. Each group explains their prediction from the data and explains this to their visitors.

10. Students may begin to practice the things they will want to say to their visitors at the Berry Feast by partnering up and walking by the various displays. Students can take turns explaining what they know about the various maps, graphs, and charts. You may even want to make arrangements for your class to partner up with another class that did not participate in these activities so that the students can practice explaining their graphs to others.

Teacher Note

We offer this extended activity for you to consider. Often, as part of the feast, gifts are given. If you have time, your students might compile and assemble a "Berry Recipe Book." Each student would ask his or her parents to copy one of their favorite berry recipes. During this activity, students practice reading and writing many new measurement words and abbreviations. The students would return the next day, share their recipes with the class, and illustrate them. Then you can make enough copies to staple together to make gift books. At the Berry Family Feast, students would be able to give one of the class recipe books to each adult visitor.

Part 2: Measuring and Cooking

Goals
- To continue exploration of standard and nonstandard measurement
- To practice estimation skills
- To practice using correct standard measures of volume
- To become familiar with math in recipes

Materials
- Several clear measuring cups
- One bag of sugar or salt or other granular ingredient
- Scrap newspapers to cover work area
- Student journals
- One large bowl (optional)

Duration
One class period

Vocabulary
Akutaq—Known as Eskimo ice cream, this is a traditional delight, or dessert, made of cloudberries, seal oil, and sugar. Now it is often made with a variety of foods and usually begins with shortening.

Preparation

In this section, students will be scooping and comparing handfuls of sugar. You may want to consider pouring the sugar into a large bowl and covering the work area with newspapers or other scraps to facilitate in the clean-up. You will want several clear measuring cups with which students can measure their handfuls. You will need a bowl of sugar, salt, or some other granular ingredient.

Instructions

1. Tell the students that they will soon be making *akutaq* for their classroom party. However, before they do, the class will need to discuss recipes and measurements used in recipes.

2. Begin discussion by asking for student volunteers to describe different times they have observed or helped someone in the family use a recipe to make something. What kinds of items did they need in order to make the recipe? Did they need to heat the food? Something to cook it in? Utensils to stir or measure with? Allow students to generate these ideas.

Activity 13: Using the Berries and Preparing for the Berry Family Feast Day 135

3. Show the empty clear measuring cups and ask if anyone has ever used a measuring cup before. As the cups are passed around from student to student, ask them to volunteer one thing that they notice about the cups. Accept all answers. Then discuss the lines on the cups, the different numbers, and the fractional units. Have students explain the meaning of the lines. Ask, "Would this be a standard or a nonstandard form of measurement? Why?" Accept student explanations.

4. Collect the cups and explain to the students that the cup is indeed a standard form of measurement, but that not everybody uses this method for measurement when they are cooking. Tell them that sometimes the Yup'ik people, and many others as well, don't even take out a measuring cup. Instead, people often use their hands and measure out what is called a "handful." Ask students if they have ever seen someone measure this way (see Figure 13.3).

Fig. 13.3: Handful of berries

5. Ask them, "How much is a handful? Can you show me?" Notice that some students may actually put both palms together. Direct them to use a single hand only. Ask for a volunteer to scoop out a handful of sugar and pour it into one of the measuring cups. Ask for two student volunteers to "read" the measuring cup and tell the class how much sugar is in the cup.

6. Ask students if they think measuring with a handful would be a standard or a nonstandard form of measurement. Have a student volunteer tally the answers on the board. This time, you reach into the bowl for a handful of sugar and pour it into a measuring cup. Have two more student volunteers "read" the measuring cup and write the answer on the board.

7. Again, ask the students what they think about measuring with a handful—is it standard or nonstandard? Allow for more examples if you feel your class needs it. Lead students to the idea that nonstandard forms of measurement work if everyone knows how much that unit should be in terms of standard units. But when they don't, a standard form of measure that everyone knows about works best. Conclude by saying that this is why recipes call for a cup of this or that, and not a pinch of this or a handful of that.

8. Have students go to their journals to draw the measuring cup they used and to show how much their "handful" measured. Watch for students who still may be struggling with this standard measure.

9. Allow students to practice the things they may want to say to their visitors at the Berry Feast about the math lessons they have been working on. A great way for students to practice is for them to partner up, or work in small groups, and practice explaining different sheets in their math journals.

Part 3: Making *Akutaq*

Goals

- To allow students the opportunity to make *akutaq*
- To continue exploration of standard volume measurements
- To identify patterns in numbers by doubling and tripling recipes
- To practice using correct measures

Materials

- Utensils and bowls, enough for each group
- Worksheet, Making *Akutaq*, one per student
- Worksheet, Sample Recipe and T-tables (recipe chosen by teacher), one per student
- Enough ingredients for the recipe of your choice (see Cultural Note, page 128)

Duration

One class period

Preparation

You will want to read through the Cultural Note regarding the history and preparation of *akutaq*. Choose the recipe(s) that you would like your students to make for the party. See the Sample Recipe and T-table worksheet we have included. White out your own copy and change the words and amounts to match the *akutaq* recipe(s) you have chosen for your class. The students will need copies of the recipe and the Making Akutaq worksheet. Make sure that you have all ingredients and utensils available. If possible, use the berries that were picked by your students or that were donated by family members to the class.

Instructions

1. Tell students that they will be making their *akutaq*, or Eskimo ice cream, for the Berry Family Feast. Pass out the Making Akutaq worksheet. Discuss the questions and write the answers the class agrees upon on the board. Student volunteers may need to display the math on the board so that everyone can agree or disagree on totals.

Fig. 13.4: Mother picking berries with her child on her back.

Activity 13: Using the Berries and Preparing for the Berry Family Feast Day

2. Point out the information boxes on the worksheet. Help students understand that one recipe will feed 20 people. Ask, "How many people would two recipes feed? What about three?"

3. Pass out the recipe sheets with the T-tables. Help students read the recipe, emphasizing each measurement and ingredient. Walk students through the first step, in which they will copy the information from the recipe.

4. Ask, "If we make two recipes, then how much of each ingredient do we need?" Tell students that they will need to break into groups to complete the process on the T-table.

5. Allow each group, as they complete their work, to share and discuss their work with other groups.

6. Have each group wash their hands, since they will be using their hands to make *akutaq*.

7. Discuss with students that they will each have to take turns using utensils while making their *akutaq*, if there are not enough utensils for everyone. They may need to discuss the best way to do this in their group. First, the students will put in the shortening. They will need to mix it extensively to fluff it up. See Linda Brown's comments in the Cultural Note to determine whether to add a little water or a little oil to each batch. Next, students will add the sugar and work the shortening back up into a fluffy texture. They will then add the applesauce, if in the recipe, and mix again. The last item to go into the *akutaq* is usually the berries. Tell students to mix their berries gently in order not to crush any.

8. *Akutaq* is best when served on the same day, or it can be kept in the refrigerator for the next day. However, if refrigerated again, it may lose its fluffy texture and its freshness.

Making *Akutaq*

Name_____

How many in our class?

_____

How many will come to our party?

_____

1 recipe	serves 20 people
2 recipes	serve ___ people
3 recipes	serve ___ peoople

How many recipes will we make?

_____

Sample Recipe and T Tables

Blueberry Akutaq

(Eskimo ice cream)

1 gallon of blueberries
1 1/2 cups of shortening
1 cup of sugar
1 cup of applesauce

Number of recipes	Number of gallons of blueberries
1	
2	
3	

Number of recipes	Number of cups of shortening
1	
2	
3	

Number of recipes	Number of cups of sugar
1	
2	
3	

Number of recipes	Number of cups of applesauce
1	
2	
3	

Activity 14: Reflections

The students have been involved in a variety of activities that were often explorative in design. This lesson provides an opportunity for open discussions regarding the past activities. For some students in Alaska, this module was their first exposure to the Yup'ik culture and a subsistence style of living. For others, the module reminded them of the yearly events in their own communities, but this may be the first time they have had math activities in their classroom based on their way of life. Use this activity as an opportunity to go back over different lessons, accepting student comments as they share their opinions and ideas with their classmates.

Goals
- To assist student memories
- To allow student documentation of learning

Materials
- *Berry Picking* storybook
- A variety of module components
- Student journals
- Math Tool Kits

Duration
Two class periods

Vocabulary
Reflections—thoughts of personal experiences

Preparation

To help spark discussion and memories of the activities they have just experienced, bring out all of the maps, posters, and any charts or graphs from this module for students to look over.

Instructions

1. Look over and discuss the pages of the *Berry Picking* storybook. Discuss favorite characters and scenes. Mention the "Mosquito Story" and the "Lazy Bear Story" and allow for responses. Ask if anyone remembered

Activity 14: Reflections

the *Big John and Little Henry* story. Ask, "Can anyone tell any of these stories from memory?"

2. Allow discussion as students share their feelings and memories of their experiences during the activities of this math module. Tell students that this kind of thought is called "reflection," and that they are going to reflect on the math activities that they have just completed. What did you like best? Was there anything funny that we did? What was the hardest activity we did? Was anything we did boring? Can anyone else think of a question we could ask each other while we reflect back on the activities?

3. After the discussion, ask students to go to their journals and write their memories of the most interesting aspects of the module. Also ask students to illustrate their story.

4. As they finish up, students may form small groups and read their stories while sharing their illustrations with each other.

5. As students complete this task, ask them to write another story about their least favorite experience. Ask them, "If there was anything you could have changed about these activities, what would it have been?" Once again, ask students to illustrate their work.

6. Allow time to share in small groups.